Developing Skills with

TABLES AND GRAPHS

Book B
Grades 6-8

Elaine C. Murphy

Contributor: Jerzy Cwirko-Godycki
Editor: Duke Hummer
Production Coordinator: Ruth Cottrell
Illustrator: John Johnson
Cover Designer: John Edeen
Compositor: WB Associates
Printer: Malloy Lithographing

ISBN 0-86651-016-8

Order Number DS01173

14 15 16 17 18-MA-99 98 97 96

DALE
SEYMOUR
PUBLICATIONS
P.O. BOX 10888
PALO ALTO, CA 94303

CONTENTS

TABLES

PICTURE GRAPHS

BAR GRAPHS

CIRCLE GRAPHS

LINE GRAPHS

REVIEW

INTRODUCTION

This is one of two books on Developing Skills with Tables and Graphs. It contains 48 worksheets that you can duplicate and use with your students for individual work or class projects. The activities in these books were designed to help students at grade levels 3-5 (Book A) and 6-8 (Book B) develop their skills in reading, using, and making tables and graphs.

ABOUT THIS BOOK

Four Major Objectives

The worksheets in this book (Book B) concentrate on tables, bar graphs, picture graphs, circle graphs, and line graphs. The exercises are designed to help you teach students

- to recognize different parts of tables and graphs such as titles, headings, and keys.
- to find specific facts in various tables and graphs.
- to make their own tables and graphs.
- to interpret and use information presented in tables and graphs.

Non-threatening Math

The exercises have been written specifically to help you focus on developing skills with tables and graphs. The size and type of numbers students encounter sometimes seems to overwhelm their thinking. In this book, students are required to use only basic arithmetic operations with which they are familiar and numbers with which they are comfortable. As a result, students should not find the computation work threatening and will be able to concentrate their efforts on learning the special skills needed to use tables and graphs effectively.

Focus on Organizational Skills

Responses to test items administered by the most recent National Assessement of Educational Progress indicate that in many cases, students' ability to read tables and graphs is superficial. They can find facts but have difficulty responding correctly to questions that require them to go beyond a direct reading of the information. In order to alleviate these problems, the activities in this book pose questions which require students to reconstruct how data in a graph has been organized, interpret the meaning of keys, and focus their attention on titles, headings, and labels. Students practice problem-solving tasks, learning to process more than one piece of information by learning to recognize whether there is enough information to answer a question, making comparisons and predictions, and drawing conclusions from facts. In addition, by studying and evaluating examples of misleading graphs, students learn what *not* to do.

USING THIS BOOK

Pick-and-Choose Lessons

Most pages are written so that individual students can work independently, giving you maximum flexibility in organizing your class. You can assign activities to individuals or small groups, or you can use them for a class lesson. You can select single exercises as needed or present a combination of worksheets as a single unit of study.

Different Kinds of Activities

There is no required order for presenting the exercises in this book. To make it easier to locate worksheets you need, the pages are organized into sections by type of graph. The first worksheet of each section presents the main features of the table or graph in that section. The other activities within a section are loosely sequenced by skills such as finding facts, using them, completing a graph or making one. For example, one page requires students to use the information from a table to decide what supplies to order for a bookstore. In another activity, students determine the best place to locate a smoke detector in a home. The last worksheet within a section suggests steps to follow when making the kind of graph studied in the section. Leading questions help students direct their efforts. Near the end of this book are two review pages, giving you a chance to check up on students' basic knowledge of tables and graphs.

Three Levels of Difficulty

To give you a better idea of what you can expect, the contents shows one, two, or three dots to indicate level of difficulty, from easiest to most difficult. These indications should be considered only as a guide because the difficulty of the exercises depends largely on the backgrounds of your students and the skills they already have acquired.

Recording Forms

Some pages suggest follow-up activities for which students must create their own graphs. Other pages may act as a natural catalyst for new ideas and graphing activities. For your convenience, two different blank forms are provided.

Beyond These Pages

Being able to read and interpret graphs, charts, and tables is an essential skill for making many everyday decisions. Draw upon different situations both in and out of math class to sharpen students' graphing skills and to point out the usefulness of presenting facts in an organized way. Expose students to a variety of graphs and tables used in real life situations — menus, TV schedules, sports statistics, and so on. Encourage students to collect their own examples and discuss the many different ways information can be organized, considering the advantages and disadvantages of each. Teach students to make and use tables and graphs to help solve problems. With time and practice, your students' skills with tables and graphs will be useful tools for making information more meaningful and manageable.

Developing Skills with

TABLES AND GRAPHS

Name ______________________________

TABLES

Facts that are related may be organized into a **table.** A table shows information in a compact form and allows you to find quickly what you want to know.

Each part of a table has a purpose.

- The **title** tells what it's about.
- The **headings** tell what kinds of facts are listed.
- The **facts** give the information and usually are listed in some kind of order.

HOW MUCH PEOPLE SLEEP IN THE U.S.A.

hours per day	number of people
less than 5	13,200,000
6	35,200,000
7	57,200,000
8	81,400,000
9	19,800,000
10 or more	13,200,000

Answer each question about the table.

1. How many people in the U.S.A. sleep 7 hours a day? ______________

HERE'S HOW TO FIND THE ANSWER:
Run your finger down the column headed **hours per day** until you find 7. Then run your finger straight across to the right.

2. How many hours do most people in the U.S.A. sleep? ______________

HERE'S HOW TO FIND THE ANSWER:
Find the greatest number in the column headed **number of people.** Then run your finger straight across to the left.

3. How many people sleep 10 or more hours? ______________

4. How many hours do the fewest number of people sleep? ______________

5. How many different items are listed in the column with the heading **hours per day**? ______________

Name ______________________

EAT AT RITA'S

Juan B. Eltee is a waiter. Part of his job is to write the order, figure the tax, and total the orders.

RITA'S MENU

SPECIALS

Hamburger	$1.50
Burrito	1.80
Fish Sandwich	1.05
Cheeseburger	1.75
Meatball Sandwich	1.65
Polish Sausage	1.60

BEVERAGES

Cola	$0.40
Coffee	0.35
Tea	0.25
Milk	0.60

SIDE ORDERS

Soup of Day	$0.60
French Fries	0.50
Bagels - plain	0.25
- with cream cheese	0.35
Cole Slaw	0.40
Guacamole	2.00
Green Salad	0.75

Sales Tax Chart

Amount	Tax
$0.00 - 0.10	none
0.11 - 0.17	0.01
0.18 - 0.34	0.02
0.35 - 0.50	0.03
0.51 - 0.67	0.04
0.68 - 0.84	0.05
0.85 - 1.10	0.06
1.11 - 1.17	0.07
1.18 - 1.34	0.08
1.35 - 1.50	0.09
1.51 - 1.67	0.10
1.68 - 1.84	0.11
1.85 - 2.10	0.12
2.11 - 2.17	0.13
2.18 - 2.34	0.14

Use the menu and the tax table to complete the following orders for Juan.

1. Hamburger ______
Cola ______
Subtotal ______
Tax ______
Total ______

2. Burrito ______
Coffee ______
Subtotal ______
Tax ______
Total ______

3. Bagel with cream cheese ______
Coffee ______
Subtotal ______
Tax ______
Total ______

4. Fish Sandwich ______
Coleslaw ______
Milk ______
Subtotal ______
Tax ______
Total ______

Name ______________________

A MARATHON WALK

Person	Spencer	Todd	Rosa	Miko	Abe	Art	Wanda	Rita	Irene	May
km walked	11	5	8	16	13	10	20	11	6	11

The school went on a marathon walk for sickle cell anemia. The table shows how far each person in one group walked.

Different stores and companies pledged money for each kilometer walked. The table shows some pledges.

Company	Spectator Sports Equipment	Off-and-On Electric Company	Cracked Comic Shop	Plastic Posies, Inc.
Pledge	$0.10	$0.20	$0.30	$0.60

Answer each question about the facts in the tables.

1. How far did Spencer walk? ______________
2. How much did Spectator Sports Equipment agree to pay? ______________
3. How much must Spectator Sports Equipment pay for Spencer? ______________
4. How much must Off-and-On Electric Company pay for Spencer? ______________
5. How much must the Cracked Comic Shop pay for Spencer? ______________
6. How much must Plastic Posies, Inc. pay for Spencer? ______________
7. What is the total contribution for Spencer's walk? ______________
8. How many kilometers were walked in all? ______________
9. How much money per kilometer was pledged in all? ______________
10. What is the total contribution for the group's walk? ______________

Name ______________________

DINNER TIME

Oscar and Priscilla are having friends for dinner.
The table shows the menu and the cooking times.

Menu	Cooking Time
Roast Chicken	$2\frac{1}{2}$ hours
Rice	40 minutes
Sweet Potatoes	30 minutes
Peas	10 minutes
Chicken Gravy	15 minutes
Chocolate Pudding	50 minutes

Answer each question about the table.

1. How long does the roast chicken take? ______________

2. How long does the rice take? ______________

3. How long does the chicken gravy take? ______________

4. Which takes the least amount of time? ______________

5. Which takes more than $\frac{1}{2}$ hour? ______________

6. Oscar and Priscilla want all the food to be ready at 7:30 P.M. At what times should they start cooking each of the following?

roast chicken ______________ peas ______________

rice ______________ chicken gravy ______________

sweet potatoes ______________ chocolate pudding ______________

7. It takes Oscar 20 minutes to get the chicken ready for roasting.
The chicken must be completely roasted by 7:30.
When should Oscar start? ______________

Name ______________________

PAULO'S PIZZA

	10" Small	12" Medium	16" Large	24" Family
Cheese only ____________	2.05	2.65	4.15	6.80
+ 1 item ____________	2.40	3.20	4.90	7.90
+ 2 items ____________	2.75	3.75	5.65	9.00
Combination ____________	3.15	4.45	6.70	10.45
cheese pizza with salami, pepperoni, olives, mushrooms, and sausage		NO SUBSTITUTIONS		
Supreme ____________	3.90	5.50	8.20	12.55
Combination with green pepper, onions, tomatoes, and extra cheese		NO SUBSTITUTIONS		
		EXTRA CHEESE add 50 cents		

You and your friends have $20 to spend.
How many of each kind of pizza can you buy?
How much change will you get? Assume there is no tax.

1. small cheese ________ ________
2. small cheese with mushrooms ________ ________
3. large cheese ________ ________
4. large cheese with pepperoni and olives ________ ________
5. medium combination ________ ________
6. family supreme ________ ________
7. medium cheese with sausage and extra cheese ________ ________
8. family combination with extra cheese ________ ________

Ingrid is having a pizza party for 18 people, including herself.
She figures that each person can eat one small pizza. Four people can eat a family pizza.

9. How many small pizzas must she order? ________________
10. How much will the pizza cost if she orders supreme? ________________
11. If Ingrid orders family pizzas, how many must she order? ________________
12. How much will they cost? ________________

Name ____________________

ORDERING SUPPLIES

Castor Hoyle works in a bookstore. One of his jobs is to order supplies. To help decide what to order for next month, Castor kept a record of sales for a week.

Item	Mon.	Tues.	Wed.	Thurs.	Fri.
pencils	14	13	14	15	12
pens	20	22	11	15	12
3-ring notebooks	1	2	4	1	0
folders	2	7	6	5	4
spiral notebooks	5	8	3	2	2
assignment books	0	0	1	1	0

1. For each of the following, find the total number of sales for the week.

pencils ________ folders ________

pens ________ spiral notebooks ________

3-ring notebooks ________ assignment books ________

2. What was the total number of sales for the week? ________

3. Suppose Castor expects 1000 items to be sold next month. That is about how many times the number of sales for the week? ________

4. If 1000 items are sold next month, predict the sales for each of the following.

pencils ________ folders ________

pens ________ spiral notebooks ________

3-ring notebooks ________ assignment books ________

Name ______________________

TIME ZONES

Some tables show patterns.
Look for patterns to help you complete the following table.

Alaska-Hawaii Time	Pacific Time	Mountain Time	Central Time	Eastern Time	Atlantic Time
12:00	2:00	3:00	4:00	5:00	6:00
1:00	3:00	4:00	5:00	6:00	7:00
2:00	4:00	5:00	6:00	7:00	8:00
3:00	5:00	______	______	______	______
4:00	______	______	______	______	______
5:00	______	______	______	______	______
6:00	______	______	______	______	______
7:00	______	______	______	______	______
8:00	______	______	______	______	______
9:00	______	______	______	______	______
10:00	______	______	______	______	______
11:00	______	______	______	______	______

Use the table to solve each problem.

1. Minneapolis is in the Central Time Zone. Boise is in the Mountain Time Zone. If it is 3:00 in Minneapolis, what time is it in Boise?

____

2. The time in Las Vegas is 3 hours earlier than the time in Charleston. Charleston is in the Eastern Time Zone. What time zone is Las Vegas in?

____

3. New York is in the Eastern Time Zone. The Rose Bowl starts at 1:00 in the Pacific Time Zone. When does a live game start on TV in New York?

____

4. San Francisco is in the Pacific Time Zone. Houston is in the Central Time Zone. If a flight leaves Houston at 9:00 and takes 3 hours and 32 minutes, when will it land in San Francisco?

____

Name ______________________

PICTURE GRAPHS

Picture graphs, sometimes called pictographs, make it easy to make comparisons between facts. They help you to see relationships quickly.

FRANKLIN MINT'S MONTHLY EXPENSES

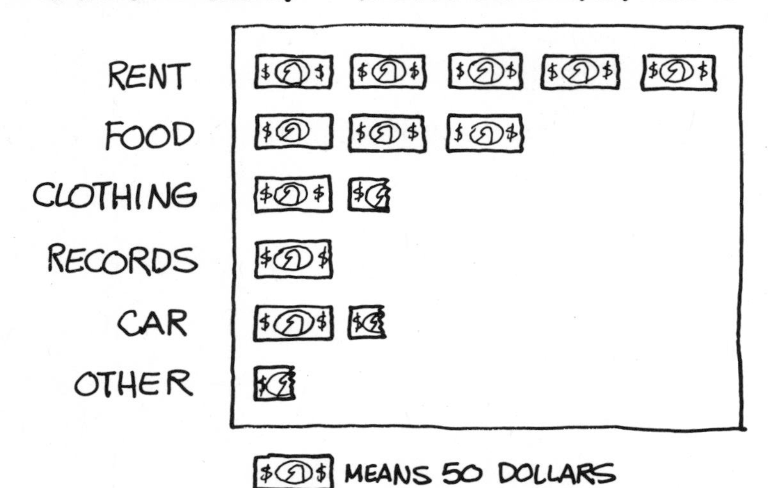

Each part of a picture graph has a purpose.

- The **title** tells what it's about.
- The **labels** tell what kinds of facts are listed.
- The **key** tells how much or how many each picture stands for.
- The **pictures** show the facts.

Take a QUICK GLANCE at the picture graph and answer these questions.

1. What is the picture graph about? ______________________
____
2. On what item does Franklin spend most of his money? ____________
3. On what item does Franklin spend the least amount of money? ____________
4. Does Franklin spend more money for clothing or records? ____________

Now take a closer look at the picture graph and answer these questions.

5. What does [$50 bill picture] stand for? ____________
6. How much money does Franklin spend for records each month? ____________
7. How much money does Franklin spend on his car each month? ____________
8. How much money does Franklin spend each month? ____________
9. Suppose Franklin wants a new sports coat that costs $150. How would you suggest he manage his money so he can pay for it?
____
____

Name ______________________

LIGHT WORK or WATT A LIFE

ENERGY USE IN THE HOME	
hair dryer	
100 watt light bulb	
electric stove	
color TV	
radio	
record player	
refrigerator	
means 100 watts per hour	

1. How much energy is used by each of the following in one hour?

   hair dryer ______________ refrigerator ______________

   color TV ______________ record player ______________

2. How much energy is used by each of the following in two hours?

   hair dryer ______________ refrigerator ______________

   color TV ______________ electric stove ______________

3. The symbol means about how many watts per hour? ______________

4. Complete the following chart.

Energy Use in the Home (measured in watts)					
hours of use	2	4	6	12	24
hair dryer					
100 watt light bulb					
electric stove					
color TV					
radio					
record player					
refrigerator					

Suppose it costs about 6 cents to use 1000 watts for one hour.

5. How much does it cost to run a color TV all day (24 hr)? ______________

6. How much does it cost to play both sides of your favorite record? ______________

Name ______________________

RAIN OR SHINE?

Robee recorded weather conditions for April.
Then he made a picture graph.

WEATHER IN APRIL

SUNNY ALL DAY	
PARTLY SUNNY, NOT RAINING	
CLOUDY, NOT RAINING	
CLOUDY, RAINY, NO THUNDERSTORMS	
PARTLY SUNNY, PARTLY RAINY, NO THUNDERSTORMS	
THUNDERSTORMS	

Each picture stands for 2 days.

Use the facts from Robee's picture graph to solve each problem.

1. Some days were sunny all day. Some days were partly sunny. How many days did Robee see the sun? ______________
2. How many days did it rain? ______________
3. How many days were cloudy? ______________
4. How many days were cloudy, but *not* rainy? ______________
5. For how many days did Robee record the weather? ______________
6. Complete this rectangular graph.

WEATHER IN APRIL

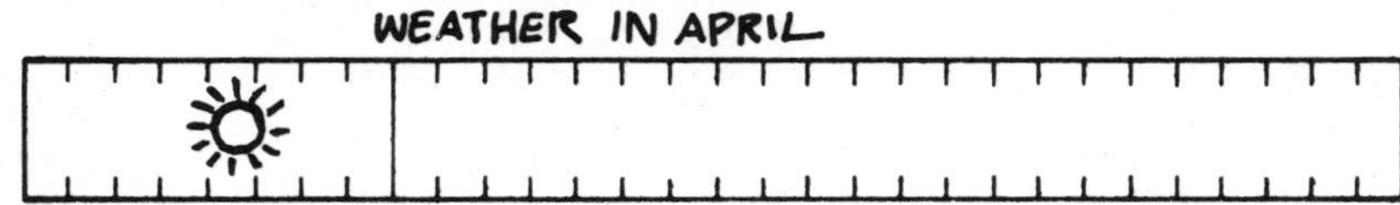

7. What fraction of the days in April were sunny? ______________
8. What fraction of the days in April were partly sunny? ______________
9. What fraction of the days in April were rainy? ______________
10. What fraction of the days in April had thunderstorms? ______________

★ Keep a record of the weather next month.
Then make a picture graph about the weather. ★

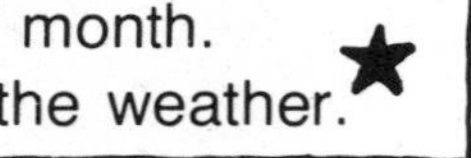

Name ____________________

STAMPS

Ellie Mint collects stamps from her penpals' letters.
This picture graph shows how many stamps she has.

STAMPS FROM DIFFERENT COUNTRIES

UNITED STATES	☐ ☐ ☐ ☐ ☐ ☐
MEXICO	☐ ☐ ☐ ☐ ☐
CANADA	☐ ☐ ☐ ▯
BRAZIL	☐
AUSTRALIA	☐ ☐ ☐ ☐
FRANCE	☐ ☐ ☐
LIBYA	☐ ▯
PHILIPPINES	☐ ☐
☐ MEANS 2 STAMPS	

Answer each question about the picture graph. Write *not enough information* if the graph does not give the answer.

1. What does ☐ mean? ____________
2. How many stamps from the United States does Ellie have? ____________
3. How many stamps from Mexico does Ellie have? ____________
4. How many stamps from Libya does Ellie have? ____________
5. How many stamps does Ellie have in all? ____________
6. How many stamps does Ellie have from Europe? ____________
7. How many stamps does Ellie have from North or South America? ____________
8. The letters from France have 3 stamps.
   How many letters did Ellie get from France? ____________
9. How many penpals does Ellie have? ____________
10. From how many different countries does Ellie have stamps? ____________

Name ____________________

RECORD RECORDING

R.J. Murphy is famous for his record collection.
This picture graph shows how many records he has.

Answer each question about the picture graph.
Write *not enough information* if the graph does not give the answer.

1. What does mean? ____________
2. What does mean? ____________
3. How many jazz records does Murphy have? ____________
4. How many country records does Murphy have? ____________
5. How many disco records does Murphy have? ____________
6. How many new wave records does Murphy have? ____________
7. How many records does Murphy have? ____________
8. How many of Murphy's records are *not* country? ____________
9. Three-fourths of Murphy's collection is records from the United States.
   How many records are from other countries? ____________
10. Which kind of music does Murphy probably like better, disco or jazz? ____________
11. Does Murphy have classical records? ____________
12. Does Murphy have movie records? ____________

Name ______________________

CENSUS SENSE

According to the 1980 census, California, Illinois, New York, Ohio, Pennsylvania, and Texas had the most people. The table at the right gives their populations.

State	Population
California	22.3 million
Illinois	11.2 million
New York	17.7 million
Ohio	10.7 million
Pennsylvania	11.8 million
Texas	13.0 million

Make a picture graph below to show the populations of these six states. Use 1 square to mean two million people.

STEP 1
Put state names here.

STEP 2
Shade squares to show how many people. Round numbers from the list to the nearest million.

Title: ______________________

________											
________											
________											
________											
________											
________											

STEP 3
Complete the key. Then make a title for the graph.

I shaded square means ______________________

Answer each of the following questions.

1. What does one-half of a shaded square mean? ____________
2. From the information in the table, which state has the greater population, Illinois or Ohio? ____________
3. What is 10.7 million rounded to the nearest million? ____________
4. What is 11.2 million rounded to the nearest million? ____________
5. How do the populations of Illinois and Ohio compare on the picture graph? ____________

Name ____________________

STEPS FOR MAKING A PICTURE GRAPH

STEP 1 FIND THE RANGE IN VALUES

What units are used? ____________

What is the greatest value? ____________

What is the least value? ____________

STEP 2 DETERMINE THE KEY

Decide what kind of picture you want to use and how big you want it.

Start with 1 picture = 1 unit. How many pictures are needed to represent the greatest value ____________

Will they fit? ____________ If not, change the key and try again.

STEP 3 LABEL THE GRAPH

Draw the key at the bottom of your graph.

How many sets of pictures will be on the graph? ____________

How much space can you allow between each set of pictures? ____________

Mark where each set of pictures starts and write their labels.

STEP 4 DRAW THE PICTURES

Use your key to determine the number of pictures for each item.

number of pictures = number of units each picture represents ÷ units per picture

Draw a light line where each set of pictures goes.

Then draw the pictures on your graph.

STEP 5 GIVE THE GRAPH A TITLE

What is your graph about? ____________________

____

Name ______________________

BAR GRAPHS

Bar graphs make it easy to compare facts. They help you to see relationships quickly.

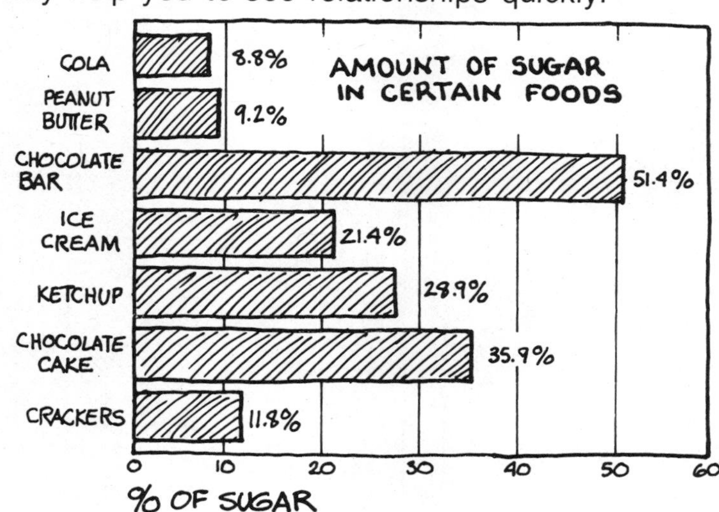

Each part of a bar graph has a purpose.

- The **title** tells what it's about.
- The **labels** tell what kinds of facts are listed and how much or how many of each.
- The **bars** show the facts.
- The **facts** give the information and usually are listed in some kind of order.

Take a QUICK GLANCE at the bar graph and answer these questions.

1. What is the bar graph about? ______________________

____

2. Which food on the graph has the greatest percentage of sugar? ______________________

3. Which food on the graph has the least percentage of sugar? ______________________

Now take a closer look at the bar graph and answer these questions.

4. Which food has a greater percentage of sugar, ketchup or ice cream? ______________________

5. What is the difference in the percentage of sugar in chocolate cake and chocolate bars? ______________________

6. How many percentage points difference is there between the food with the greatest percent of sugar and the food with the least? ______________________

7. List the foods according to the percent of sugar, from least to greatest.

____

8. The labels across the side of the graph represent various percents. They start at what number and end at what number? ______________________

Name ______________________

GRAPHIC FACTS

In each problem, only one graph shows *all* the facts correctly.
Circle the correct graph. Draw an X on each of the other two graphs.

1. Mr. McIntosh's class voted on their favorite fruit. Each student voted once.
Here is the vote: bananas, 3   grapes, 2   oranges, 6
apples, 5   strawberries, 4   pears, 1

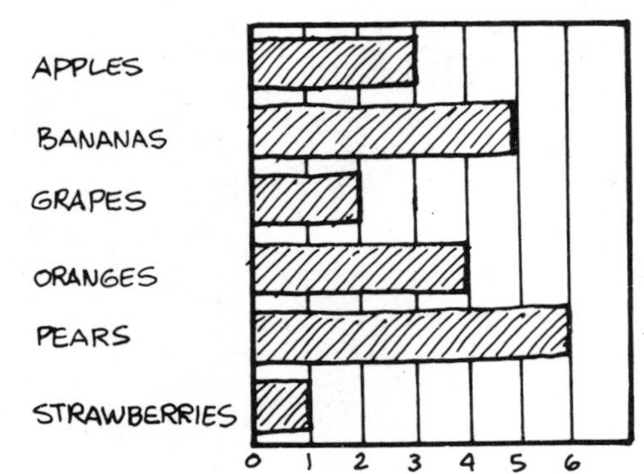

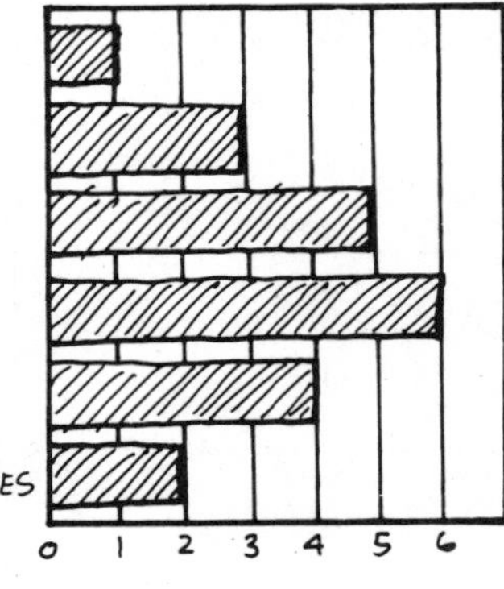

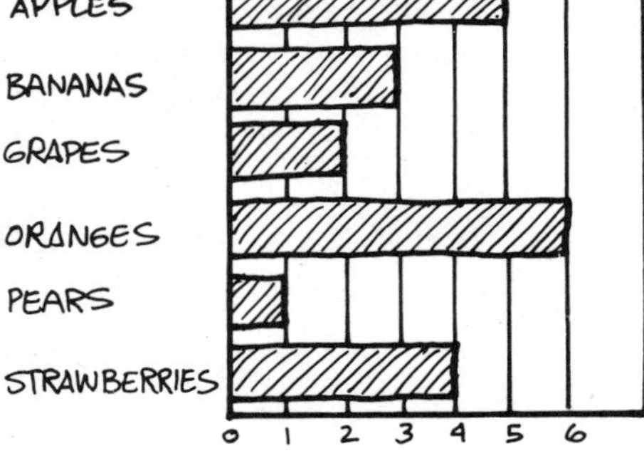

2. Ms. Mallard's class voted on the ugliest animal. Each student voted once.
Here is the vote: skunk, 4   bat, 4   gopher, 3
rat, 6   armadillo, 5   porpoise, 1

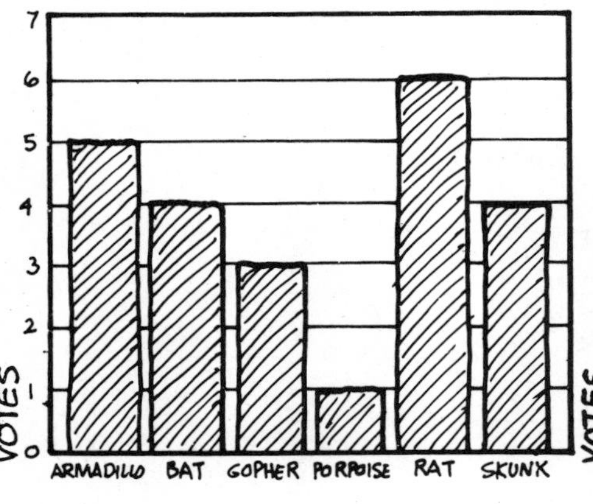

7 6 5 4 3 2 1 0
VOTES
ARMADILLO BAT GOPHER PORPOISE RAT SKUNK

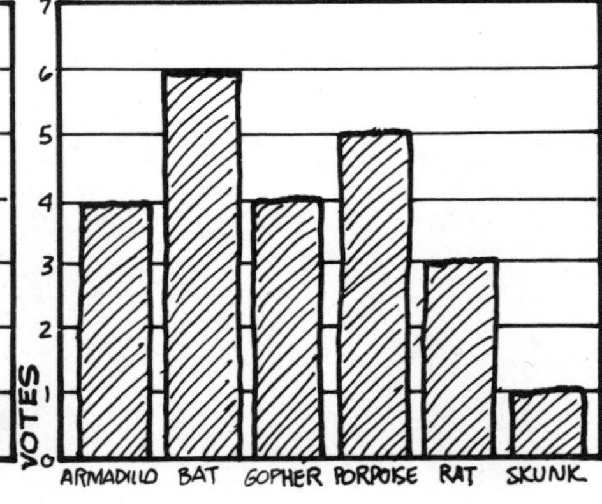

3. The ages of 7 trumpet players in band are 13, 12, 11, 12, 11, 10, 12.

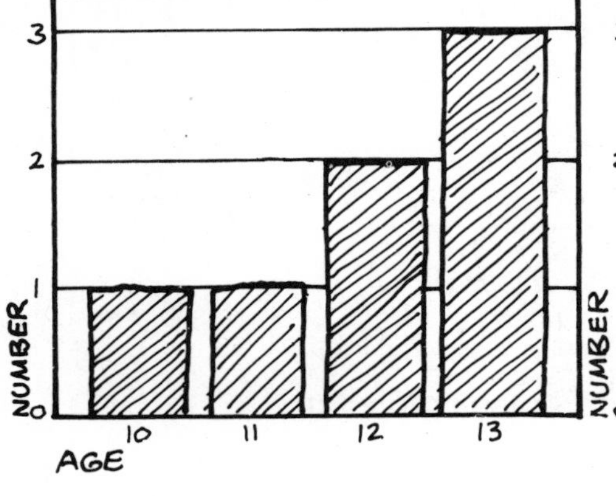

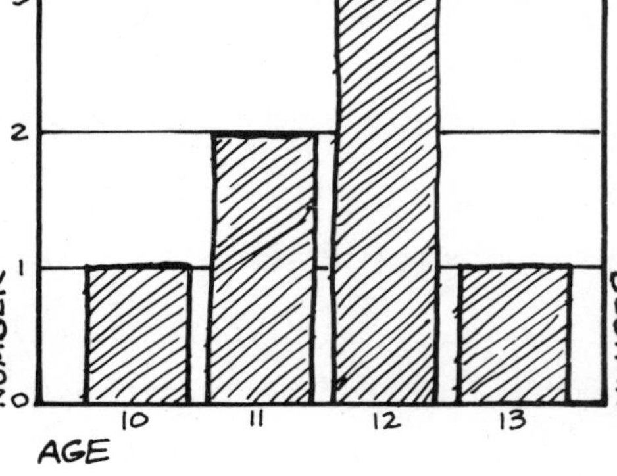

Name ______________________

TRICKS OF THE TRADE

Graphs can be misleading. For example, this graph seems to show facts without actually giving any information. It is misleading because it has no labels on the left-hand side.

MIRACLE PLUS★ GIVES YOU MORE!

BRAND X
BRAND Y
MIRACLE PLUS★
BRAND Z

This list gives some reasons why different graphs can be misleading.

- no labels or some labels missing
- measurements in scale not consistent
- scale does not start at 0
- scale is too spread out, exaggerating differences
- width of bars not the same

Explain why each of the following graphs is misleading.

1. ______________________
____

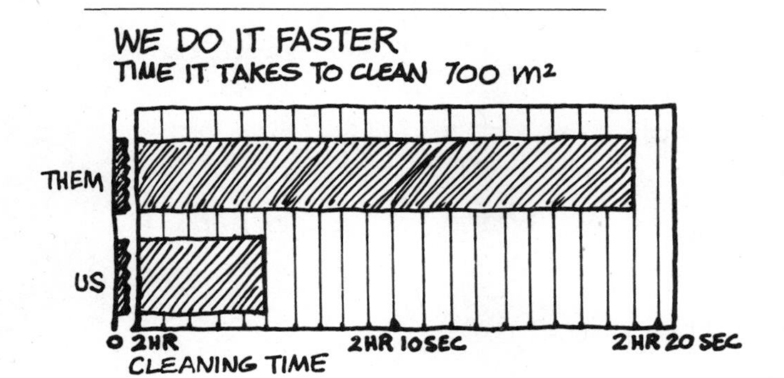

2. ______________________
____

3. ______________________
____

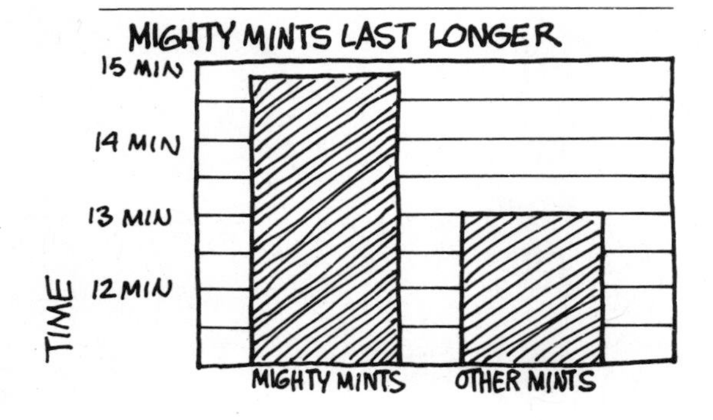

4. ______________________
____

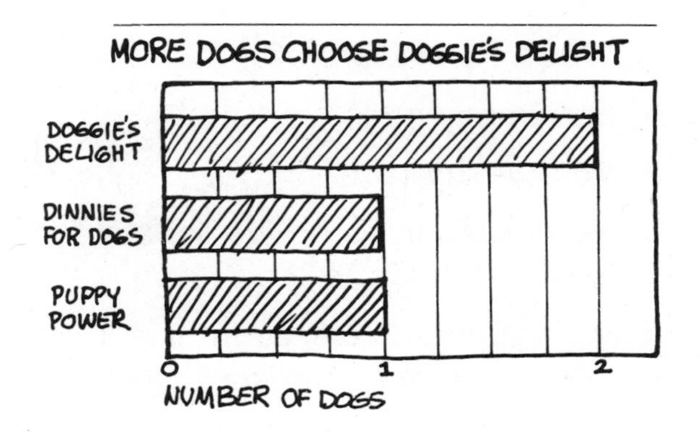

Name ______________________

U.S. GIANTS

Both the table and the graph show facts about giant trees.

Giant Trees in the U.S.

Tree and Location	Height
Acacia, HA	43 m
Bold Cypress, NC	42 m
American Birch, MI	49 m
Painted Buckeye, GA	44 m
Port-Oxford Cedar, OR	67 m
Coast Redwood, CA	110 m

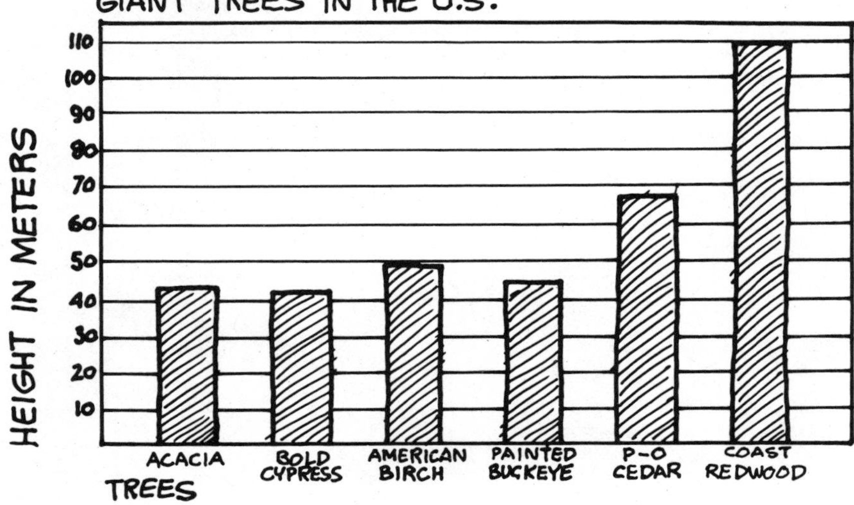

Some questions are easier to answer using the table.
Other questions are easier to answer using the graph.
Check which you would use. Use a ✓.

	Table	Graph	Either
1. How tall is the Bold Cypress?	☐	☐	☐
2. Which is taller, the Painted Buckeye or the American Birch?	☐	☐	☐
3. Where is the Port-Oxford Cedar located?	☐	☐	☐
4. Which tree is tallest?	☐	☐	☐
5. How many trees are listed?	☐	☐	☐
6. What is the range of tree heights?	☐	☐	☐
7. How many trees on the list are less than 60 m tall?	☐	☐	☐
8. What is the difference in height between the Acacia and the Coast Redwood?	☐	☐	☐

Name ______________________

WHAT ARE YOU AFRAID OF?

Three thousand people in the U.S. were asked, "What are you the most afraid of?" The following graph shows the results. Some people named more than one fear.

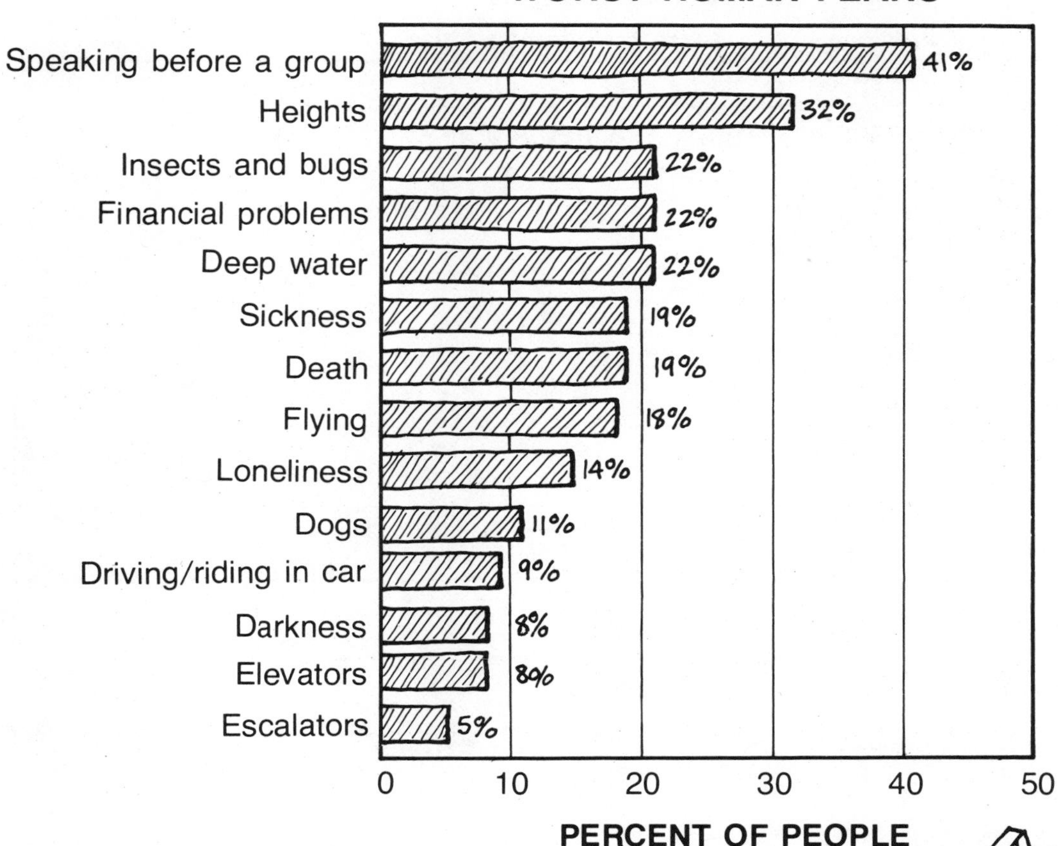

Which conclusions can you make from the graph?
Write *yes* if you can, *no* if you cannot.

_______ **1.** 11% of all dogs are afraid.

_______ **2.** People are more afraid of loneliness than darkness.

_______ **3.** Many people think bugs are frightening.

_______ **4.** Deep water is more frightening than darkness.

_______ **5.** Most people like flying.

_______ **6.** Most people would rather be alone than speak before a group.

_______ **7.** More people named heights than named sickness.

_______ **8.** The same people who named insects and bugs named deep water.

Name ______________________

ON AND OFF THE BUS

The following chart shows how many riders got on and off the number 32 bus on one run.

BUS STOP	Alder St.	Birch St.	Elm St.	Hickory St.	Larch St.	Maple St.	Oak St.	Spruce St.
PEOPLE ON	8	4	9	3	11	0	7	2
PEOPLE OFF	0	4	0	2	5	7	8	4

Suppose there were 5 people on the bus when it reached Alder Street. Complete the following bar graph showing how many people were on the bus between stops.

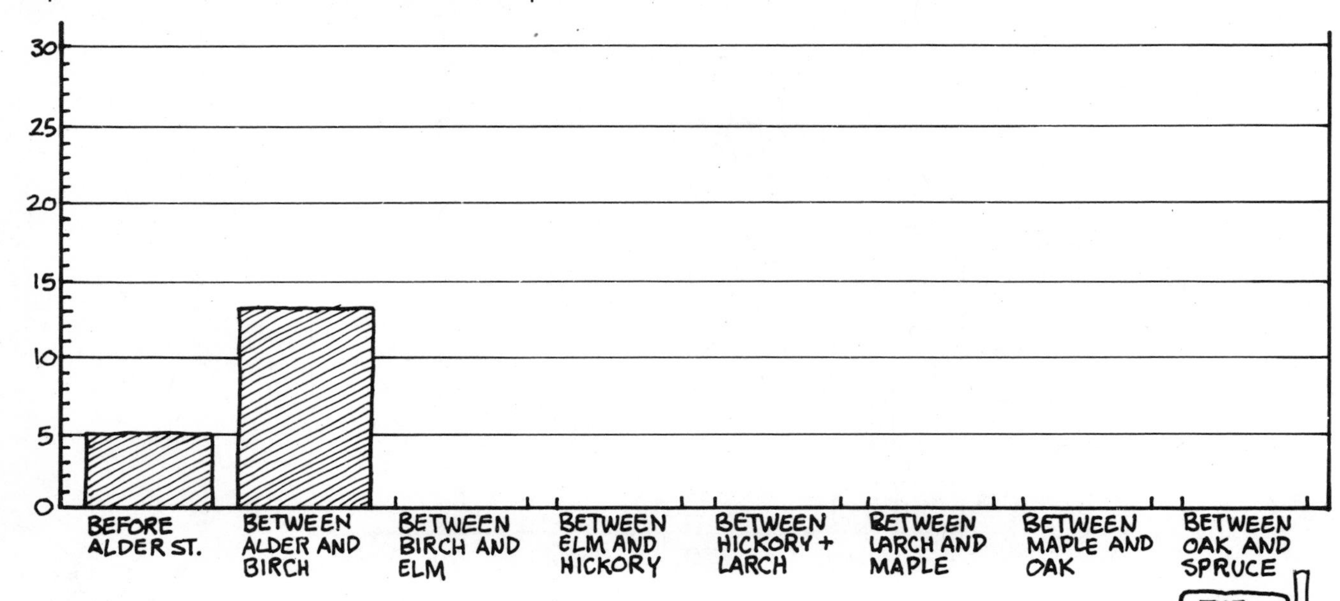

1. When did the bus have the most people? ______________

2. When did the bus have the fewest people? ______________

3. Were there more people between Hickory St. and Larch St. or between Larch St. and Maple St? ______________

4. How many people were on the bus after Oak Street? ________

A new bus stop will be added to the number 32 route. Use the information from the table to decide where you think the new bus stop should be. Why?

____

____

Name ______________________________

HOW TALL ARE YOU?

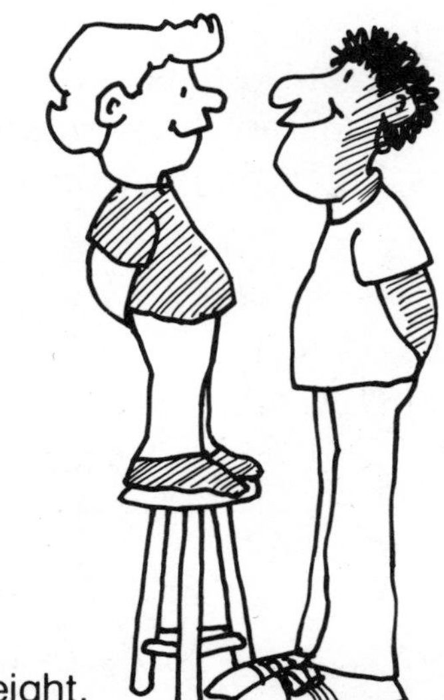

The kids in Harlan's class measured their heights in centimeters. Here are their measurements.

97 cm, 101 cm, 103 cm, 106 cm, 108 cm,
103 cm, 99 cm, 101 cm, 102 cm, 104 cm,
103 cm, 105 cm, 105 cm, 102 cm, 102 cm,
105 cm, 100 cm, 103 cm, 105 cm, 103 cm,
104 cm, 106 cm, 101 cm, 103 cm, 101 cm,
103 cm, 105 cm, 103 cm, 105 cm, 104 cm,

Harlan's job is to make a bar graph about the heights. Complete the graph for Harlan. Use these hints to help.

1. Make a tally below showing the number of kids for each height.

2. What is the greatest number of kids for a height? ______________

   Use this number to decide what numbers go along the side.

3. What is the shortest height? ______________

   What is the tallest height? ______________

   Use these numbers to decide what numbers go along the bottom.

4. Draw a bar for each height to show how many kids.

5. Make a title for your graph.

Title: ______________________________

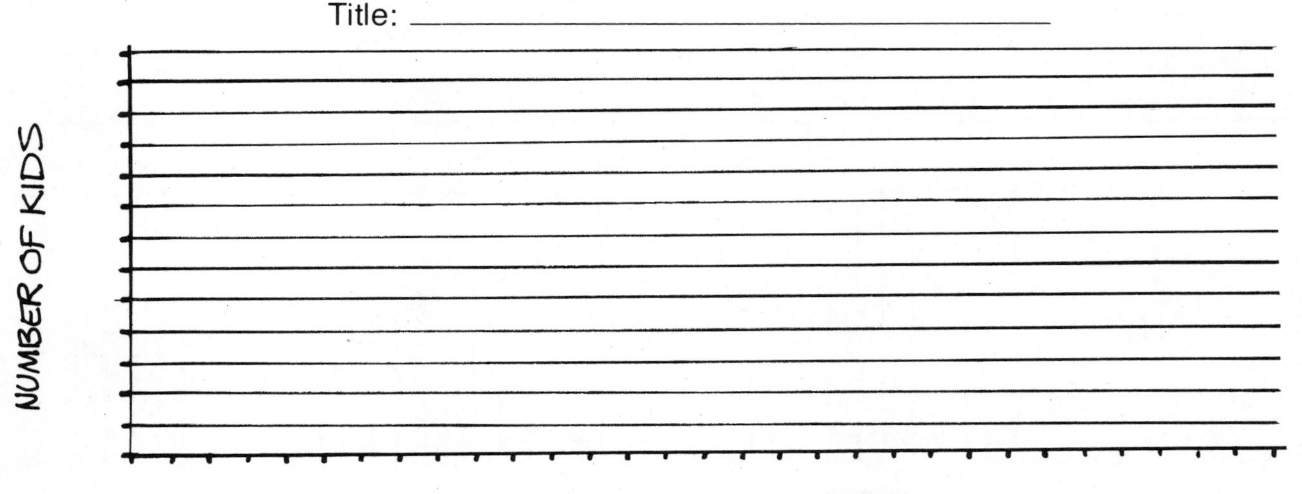

Name ____________________

DICE DETECTOR

Four different people have their own dice. They each rolled their dice 144 times. Graphs were made to show the results. Match the correct names to the graphs. Use these clues.

- Anita Win rolled more sevens than anyone else.
- Willy Maykett had 4 twelves.
- Bette A. Million had 5 more tens than elevens.
- Ben D. Rules has unfair dice.

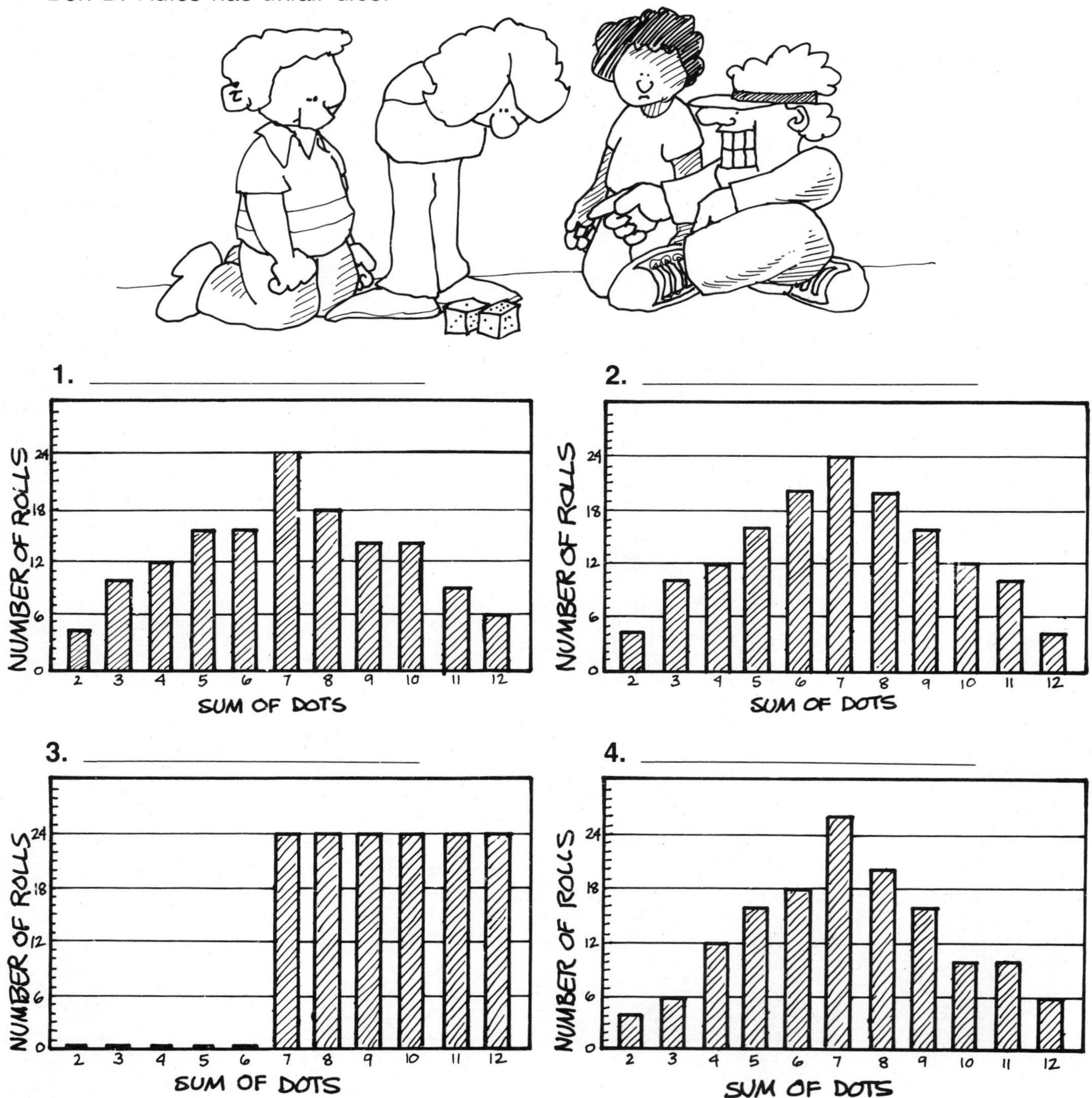

Name ____________________

DARTS

Juana Win was practicing darts with this board. She had 5 darts to throw. Here is a list of her scores.

30	32	35	40	30	22
35	46	42	45	40	35
28	34	38	38	36	32

1. Complete the tally to show how many scores

Score	Number
20 – 25	
26 – 30	
31 – 35	
36 – 40	
41 – 45	
46 – 50	

2. Use your tally to complete a bar graph.

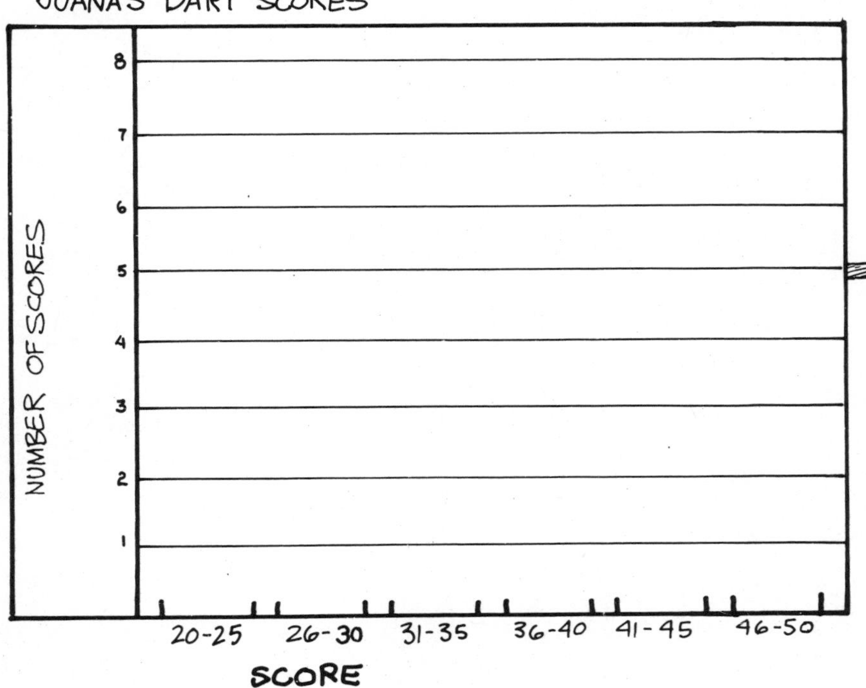

3. If Juana plays against someone who usually scores about 45, do you think Juana will win? ________

Name ______________________

STEPS FOR MAKING A BAR GRAPH

STEP 1 FIND THE RANGE IN VALUES.

What units are used? ______________

What is the greatest value? ______________

What is the least value? ______________

STEP 2 DETERMINE A SCALE.

Start with 1 cm = 1 unit. What is the length of the longest bar? ______________ Will it fit? _____

If not, change the scale and try again.

STEP 3 LABEL THE GRAPH.

Mark each centimeter along the side of the graph.

Label the marks by the units they represent.

Then, decide how wide each bar should be.

How many bars will be on the graph? ______________

How much space can you allow between each bar? ______________

STEP 4 DRAW THE BARS.

Mark where each bar starts and write the labels.

Use your scale to determine the length of each bar.

Then, draw the bars on your graph.

Check two close, but different values.

Will their bars show a difference? ______________

bar length in centimeters = number of units bar represents ÷ units per centimeter

STEP 5 GIVE THE GRAPH A TITLE.

What is your graph about? ______________________________

____

Name ______________________

CIRCLE GRAPHS

Circle graphs help show how parts of something are related to the whole thing. For example, this circle graph shows how Jolene Simpson spends each part of her earnings.

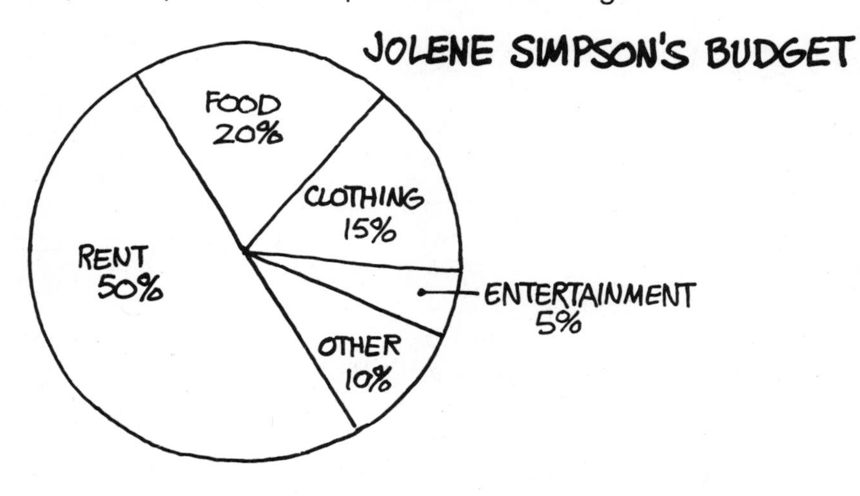

Each part of a circle graph has a purpose.

- The **title** tells what it's about.
- The **labels** on the circle tell the facts.
- The **sections** of the circle show what fraction or part of the whole the facts represent.

Take a QUICK GLANCE at the circle graph and answer these questions.

1. What is the circle graph about? ______________________

____

2. How many different ways does Jolene spend her earnings? __________

3. How does Jolene spend most of her money? __________

Now take a closer look at the circle graph and answer these questions.

4. How many cents out of each dollar Jolene earns go for the following?

food __________ clothing __________ other __________

rent __________ entertainment __________

5. Suppose Jolene earns $400 each month (after taxes).

What does she pay for rent? __________

6. If Jolene pays $250 each month for rent,

how much does she earn? __________

Name ____________________

UNDER A SPELL

In each problem, only one graph shows *all* the facts correctly.
Circle the correct graph. Draw an X on each of the other two gaphs.

1. When asked to spell **dinosaur,** 25 people spelled it correctly, 60 spelled it incorrectly, and 15 refused to try.

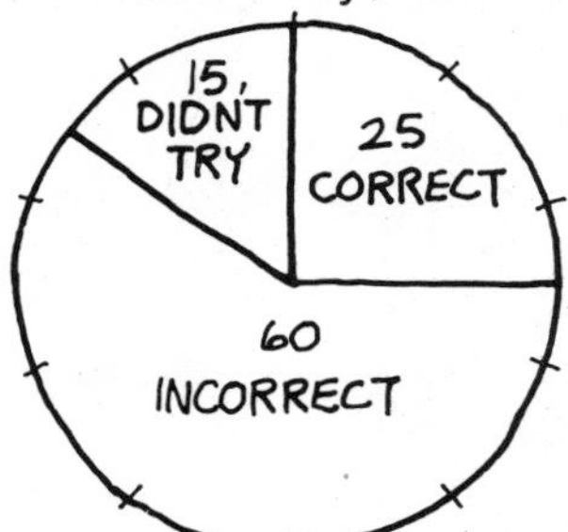

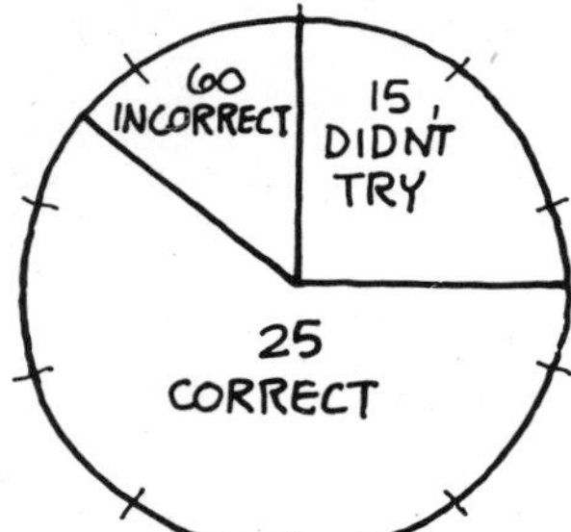

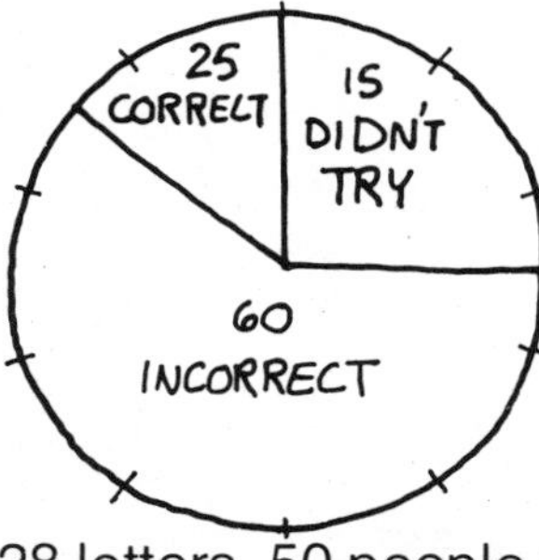

2. When asked if **antidisestablishmentarianism** has 28 letters, 50 people said *yes,* 35 people said *no,* and 15 people *didn't know.*

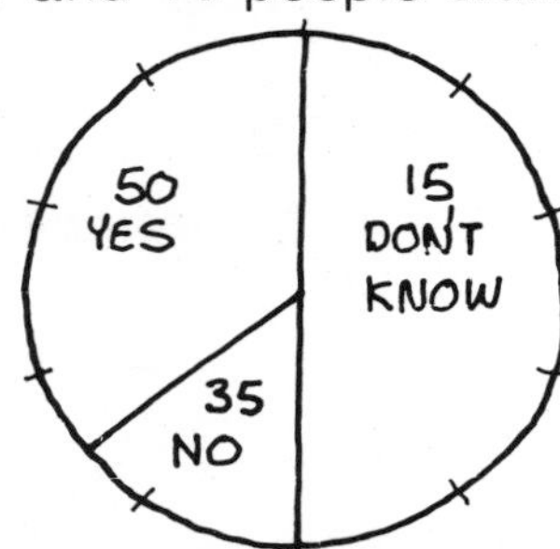

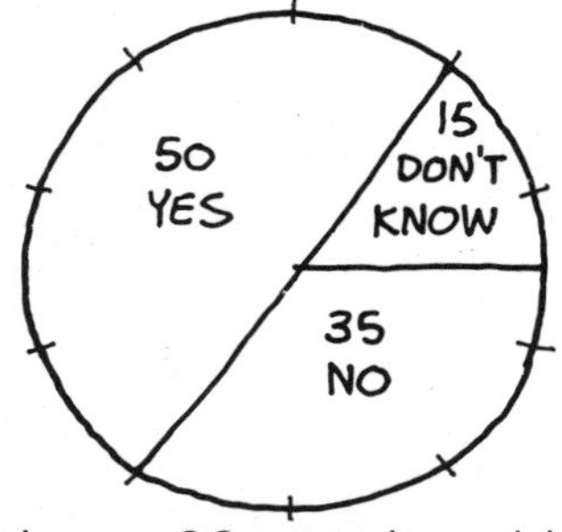

3. When asked, "Can you spell well?" 30 people said *yes,* 30 people said *no,* and 40 people *didn't know.*

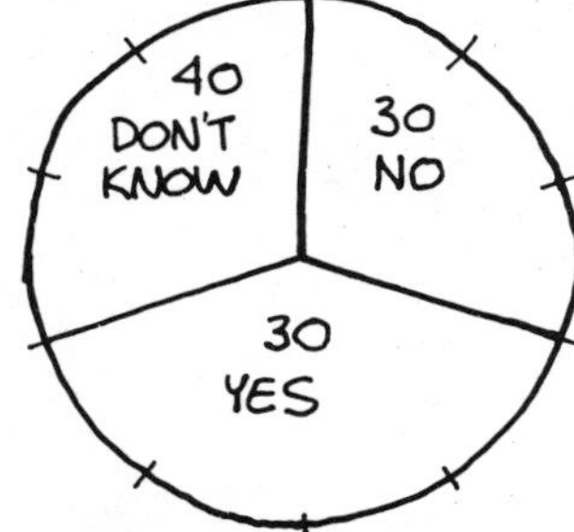

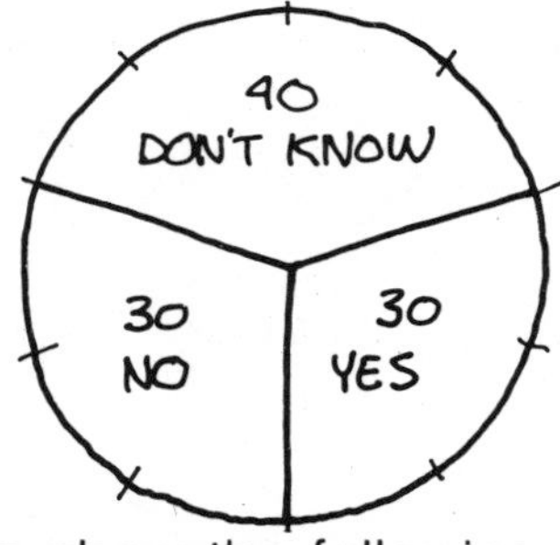

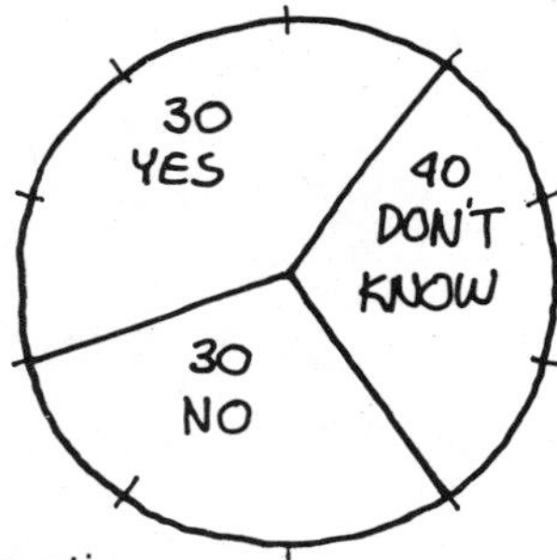

Label the graph correctly to show the following information.

4. When asked to spell **encyclopedia,** 20 people spelled it correctly, 50 people spelled it incorrectly, and 30 people refused to try.

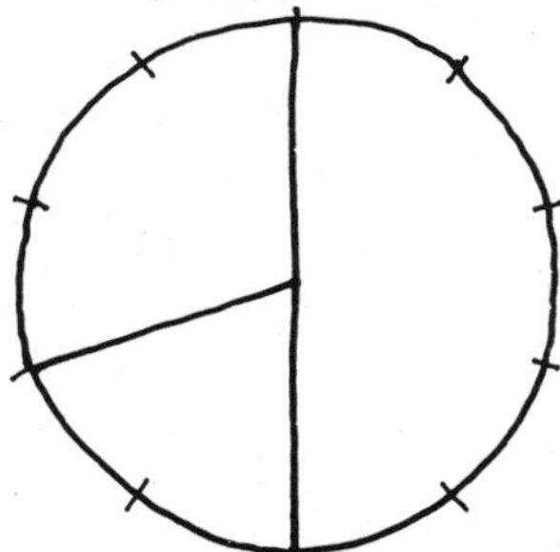

Name ______________________

HOUSE PLANTS

Lynette asked 25 people about the plants in their homes. Some people had more than one plant. She counted 85 plants in all.

NUMBER OF PLANTS

rubber plants	HHT HHT I
snake plants	HHT III
geraniums	HHT HHT III
begonias	HHT HHT
cacti	HHT HHT HHT I
jade plants	HHT I
palm trees	HHT HHT I
other	HHT HHT

1. Lynette started a chart about the plants. Finish Lynette's chart. The decimals are rounded to the nearest hundredth.

plants	fraction	decimal	percent
rubber plants			
snake plants			
geraniums			
begonias			
cacti			
jade plants			
palm trees			
other			

2. Lynette drew a circle graph with sections for each type of plant. Finish labeling the circle graph.

3. Rank the plants in order from most popular to least popular.

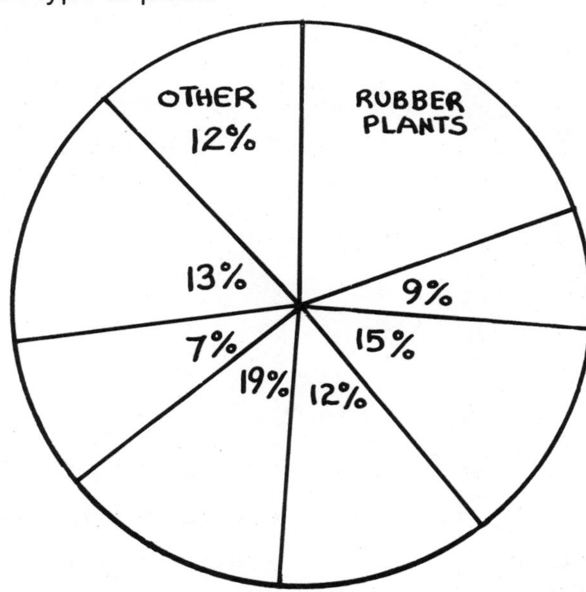

(1) ______________ (5) ______________

(2) ______________ (6) ______________

(3) ______________ (7) ______________

(4) ______________ (8) ______________

Name ______________________

BE CAREFUL!

Home fires start in different places. For every 100 fires, the picture shows where they are likely to start.

1. Complete the circle graph about home fires. Give the graph a title and labels. Group all sources less than 10 fires into a category called OTHER.

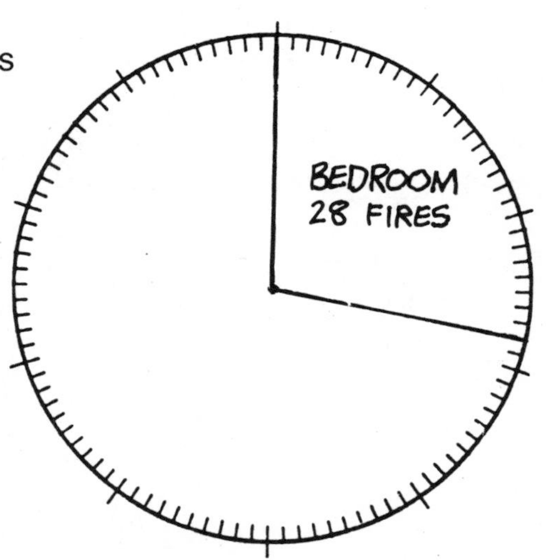

2. Where do most home fires occur? ______________

3. Out of 100 fires, how many start in the bathroom? ______________

4. What are the two best places for smoke detectors in a home?

____

Name ______________________

THE EYES HAVE IT

Professor Jackson Neville is studying the students at Almost High School. This table shows what he found in one class.

Student's Hair and Eyes

eye color	number with light hair	number with dark hair
brown	5	13
blue	9	7

Use the information from the table to complete each of the following.

	How many?	About what percent of total?
1. total	34	100 %
2. light hair	14	41 %
3. dark hair	20	____%
4. brown eyes	____	____%
5. blue eyes	____	____%
6. light hair and brown eyes	____	____%
7. light hair and blue eyes	____	____%
8. dark hair and brown eyes	____	____%
9. dark hair and blue eyes	____	____%

Professor Neville's assistant, Imogene Yus, started a graph about the results.

10. Complete the graph. Include a title.

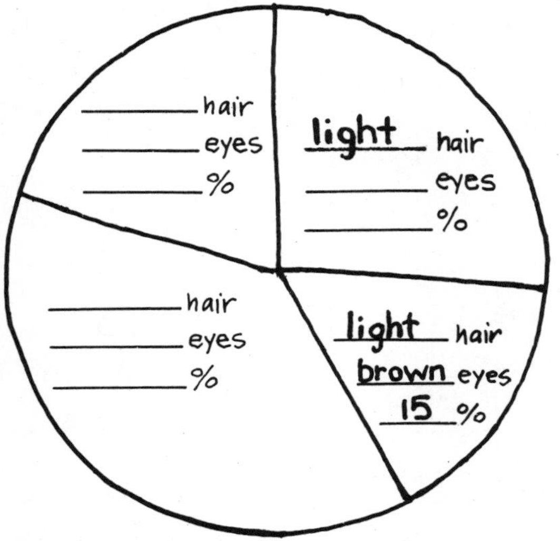

Suppose there are 2000 students in Almost High School.
Predict the number of students who have the following.

11. light hair & brown eyes ______

12. light hair & blue eyes ______

13. dark hair & brown eyes ______

14. dark hair & blue eyes ______

Name ______________________

TV SHOWS

Terry Dacktal made a record of the TV shows for one evening from 4 PM to 4 AM. He counted the different kind of shows that were on each half hour. Then he made a circle graph to show the results.

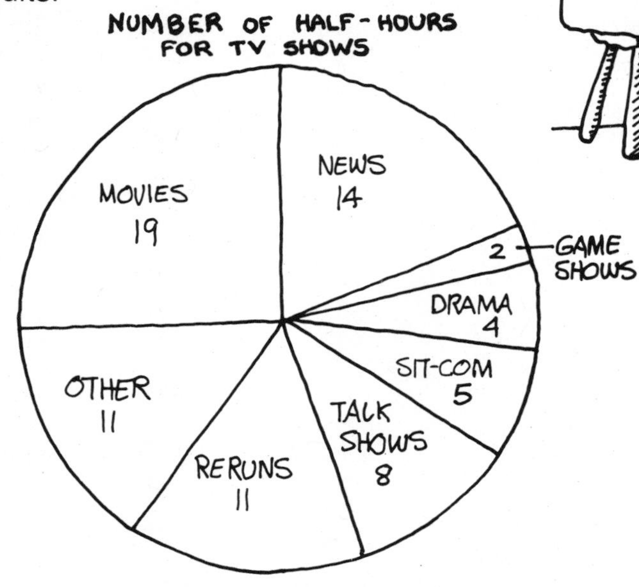

1. What is the title of Terry's graph?

____

2. What are the labels on the graph?

____

3. How many different kinds of shows did Terry record? ____________
4. How many half-hours of news were counted? ____________
5. How many half-hours of game shows were counted? ____________
6. Which kind of show was on for the most amount of time? ____________
7. Which kind of show was on for the least amount of time? ____________
8. Did Terry count more time for sit-coms or drama? ____________
9. Do you think Terry made his record about a Saturday, a Sunday, or a weekday? If you're not sure, how could you find out?

____

Use a newspaper or TV guide to count the amount of time for different TV shows for one whole day. Draw a circle graph to show your results. How is your graph different from Terry's?

Name ______________________

COST OF LIVING

Willy Maykett and Kenny Maykett share an apartment. They split the costs shown on the circle graph in half.

1. How much does Willy pay each month for each of the following?

   rent ______________

   gas ______________

   electric ______________

   water ______________

   rent and utilities ______________

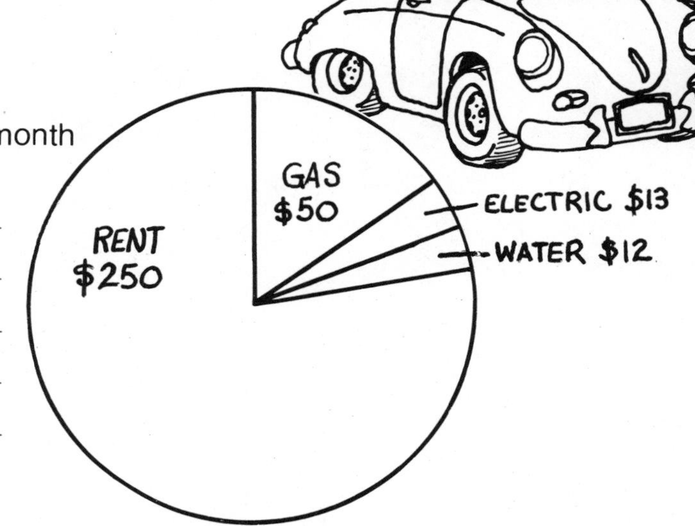

The circle graph below shows how Kenny spent his earnings last year.

2. About how much did Kenny spend each month for each of the following?

   rent and utilities ______________

   food ______________

   car ______________

   clothing ______________

   entertainment ______________

   other ______________

   total ______________

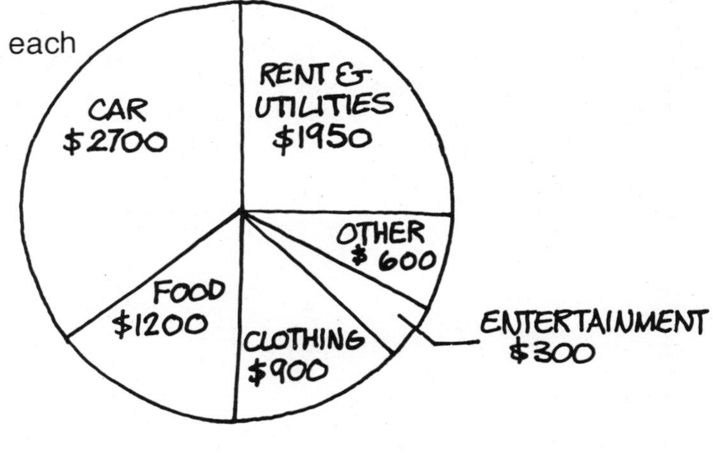

3. Last year Kenny's car expenses were as follows.

   payment, $1200

   insurance, $450

   registration & fees, $100

   parking and tolls, $250

   gas, $500

   repairs, $200

   Make a circle graph to show Kenny's car expenses.

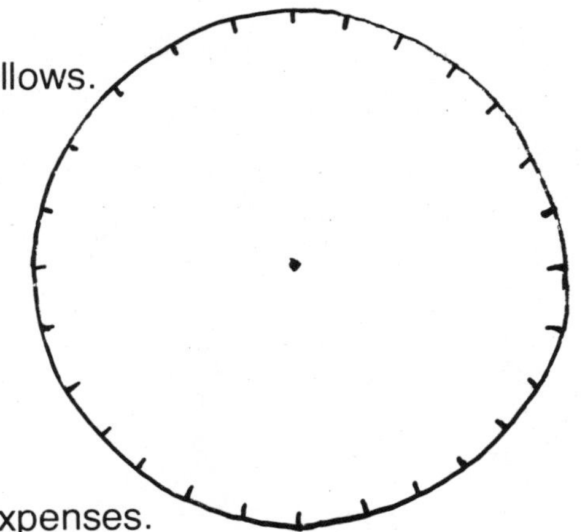

Name ______________________

STEPS FOR MAKING A CIRCLE GRAPH

STEP 1 FIND THE WHOLE

The entire circle represents 100% of something such as total expenses or the total number of items.

What is the subject of your graph? ______________

What is the total value for the items on your graph? ______________

STEP 2 FIND THE PARTS

Each item to be graphed represents a part of the whole. To complete the circle graph, you must find exactly what fraction or percent each item represents.

FRACTION: $\frac{\text{part}}{\text{whole}}$ PERCENT: $(\text{part} \div \text{whole}) \times 100$

STEP 3 FIND THE DEGREES FOR EACH PART

Every circle is made up of 360°. Use the following equation to find the angle measure for each item.

$$\text{angle measure for item} = 360 \times \text{fraction or percent of whole item}$$

Check your work. Do the angle measures total 360 in all? ______________

STEP 4 DRAW AND LABEL THE PARTS

Draw a circle and a radius. Then use a protractor to draw each angle.

Where will you place the center of the protractor? ______________

Where will you line up the protractor? ______________

Where is the first line segment drawn? ______________

Each new angle is measured from the previously drawn line segment.

How will you line up the protractor for the next angle? ______________

Label each part of your graph.

STEP 5 GIVE THE GRAPH A TITLE

What is your graph about? ______________________________

____

Name ______________________

LINE GRAPHS

Line graphs use dots connected by lines to show facts.
The position of the lines makes it easy to see how things change.

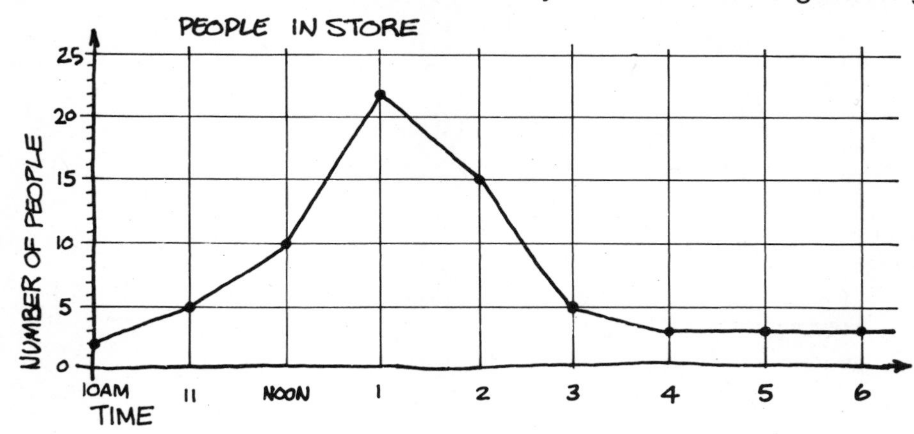

Each part of a line graph has a purpose.

- The **title** tells what it's about.
- The **labels** along the sides and across the bottom tell what kinds of facts are listed.
- The **scales** along the sides and across the bottom are used to tell how much or how many.
- The dots or **points** show the facts.
- The **lines** connecting the points give estimates of values between the points.

Take a QUICK GLANCE at the bar graph and answer these questions.

1. What is the line graph about? ______________________

____

2. What is the busiest time of day at the store? ____________

3. About what time of day does business slow down? ____________

Now take a closer look at the line graph and answer these questions.

4. The store opens at 10 AM. How many people are in the store when it opens?

____

5. About how many people are in the store at 2:30 PM? ____________

6. What was the greatest number of people in the store at any one time?

____

7. What was the least number of people in the store at any one time?

____

Name ______________________

WHEN IT RAINS IT POURS

Line graphs help show changes. Each shape below is from a line graph that shows the change in rainfall or snowfall for a city in the table. Use the facts from the table to name each shape by its city.

1. ______________________

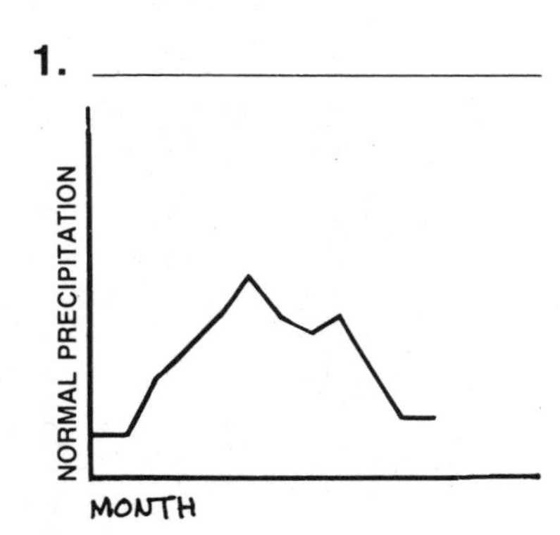

2. ______________________

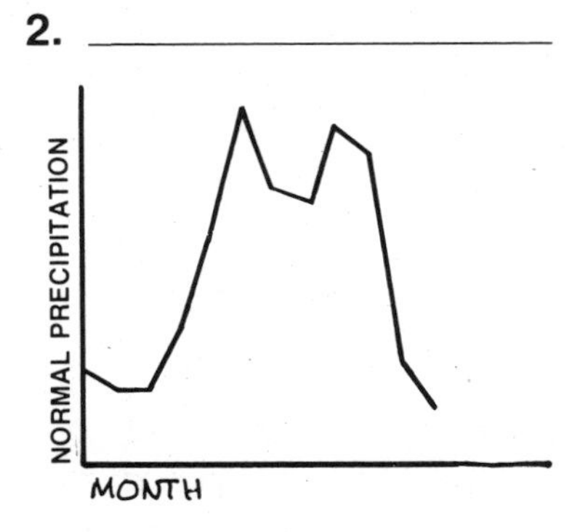

3. ______________________

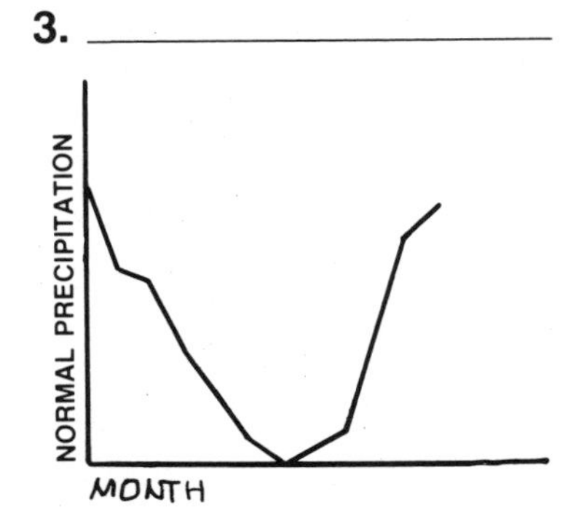

4. ______________________

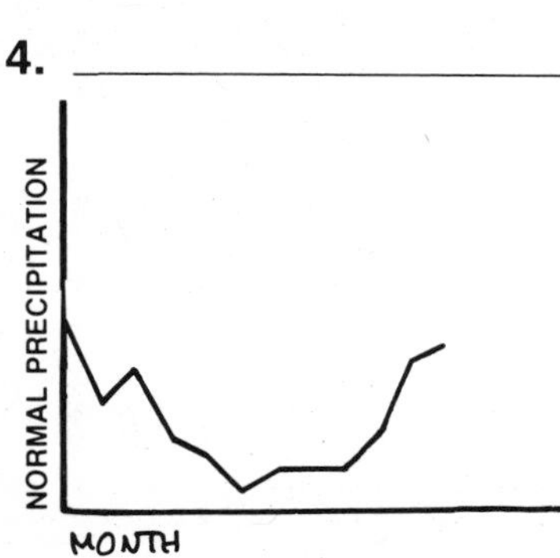

5. ______________________

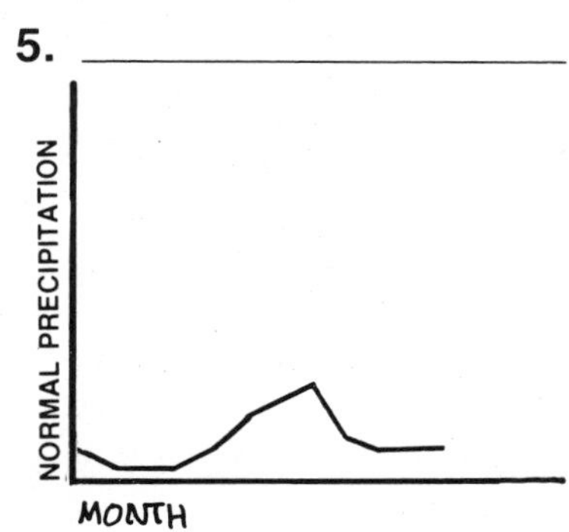

6. ______________________

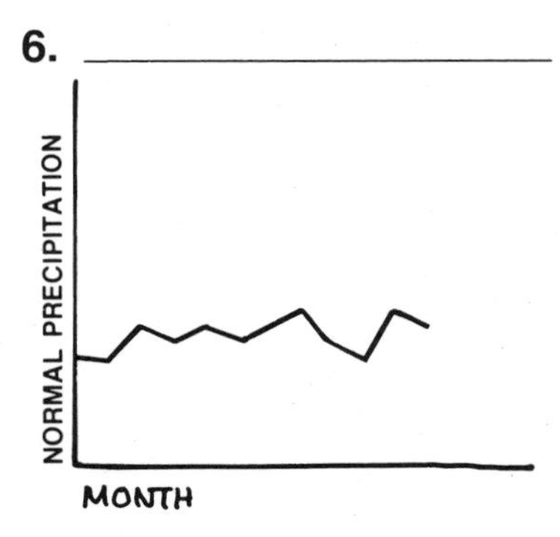

NORMAL PRECIPITATION/SNOWFALL

(in centimeters)

	Miami	Eureka	Honolulu	Fairbanks	New York	Kansas City
Jan.	6	19	11	2	7	3
Feb.	5	13	6	1	7	3
Mar.	5	12	8	1	9	7
Apr.	9	8	4	1	8	9
May	15	5	3	2	9	11
Jun.	23	2	1	4	8	14
Jul.	18	0	2	5	9	11
Aug.	17	1	2	6	10	10
Sep.	22	2	2	3	8	11
Oct.	21	8	4	2	7	8
Nov.	7	15	8	2	10	4
Dec.	4	17	9	2	9	4

Name ______________________

RUNNING AWAY

The line graph below shows winning times for the women's Olympic 100 meter run.

1. What was the winning time in 1960? ______________
2. What was the winning time in 1964? ______________
3. How did the winning times change?

   Write *increased, decreased,* or *stayed the same* on the blanks.

   from 1948-1952 ________________ from 1964-1968 ________________

   from 1952-1956 ________________ from 1968-1972 ________________

   from 1956-1960 ________________ from 1972-1976 ________________

   from 1960-1964 ________________

4. The graph does *not* show the winning time in the 1980 Olympics. Based on this graph, which time was most likely the winning time in 1980? Circle it.

   11.70 sec 10.70 sec 9.70 sec

   Find the actual 1980 time in an information almanac and compare it to your prediction.

Use an information almanac to find times for the men's Olympic 100 meter run. Make a line graph to show how the times changed from year to year. Try to guess the 1980 result before you read it.

Name ____________________

THE KING'S TOSS

Rufus Lee King tossed a coin 100 times.
The chart below shows his results.

1. Complete the chart.

NUMBER OF HEADS	7	11	18	24	32	36	42	44	45	50
NUMBER OF TOSSES	10	20	30	40	50	60	70	80	90	100
HEADS ÷ TOSSES	0.70	0.55	0.60							

2. Use the information from the chart to complete the line graph.

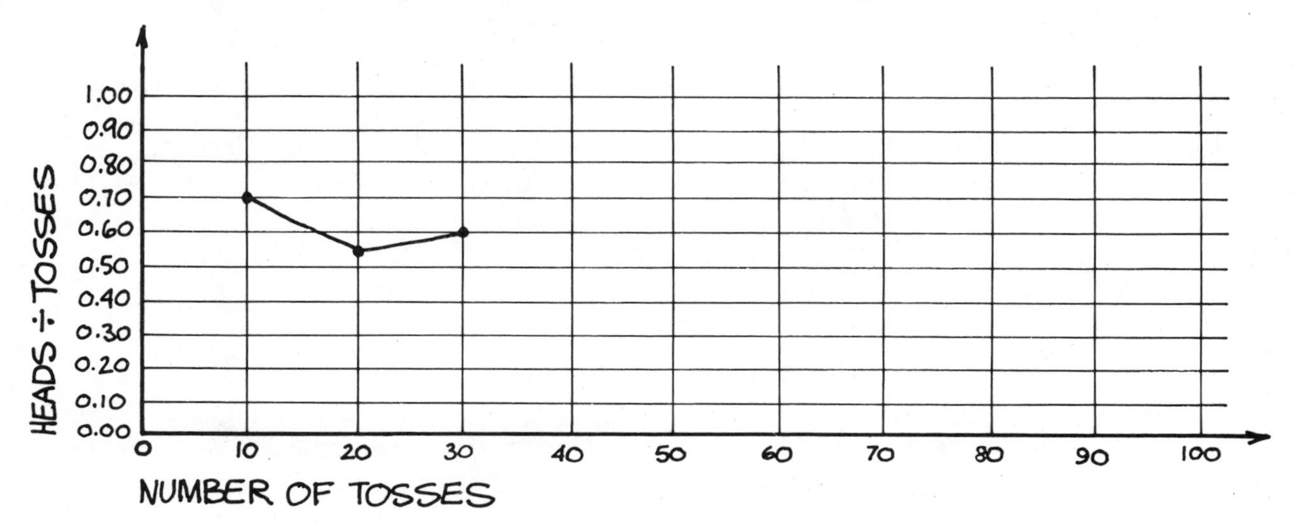

3. How does the graph change? Write *increases, decreases,* or *stays the same* on the blanks.

from 10 tosses to 20 tosses ____________

from 20 tosses to 30 tosses ____________

from 30 tosses to 40 tosses ____________

from 40 tosses to 50 tosses ____________

from 50 tosses to 60 tosses ____________

from 60 tosses to 70 tosses ____________

from 70 tosses to 80 tosses ____________

from 80 tosses to 90 tosses ____________

from 90 tosses to 100 tosses ____________

4. If Rufus tossed the coin 10 more times (110 times in all), what number would you expect for **heads ÷ tosses**? ____________

5. If Rufus tossed the coin 10 more times (110 tosses in all), how many heads would you expect him to see altogether? ____________

Name ____________________

TRADE SECRETS

MOISHE'S MARBLES wanted to show how its sales had improved. It used this line graph.

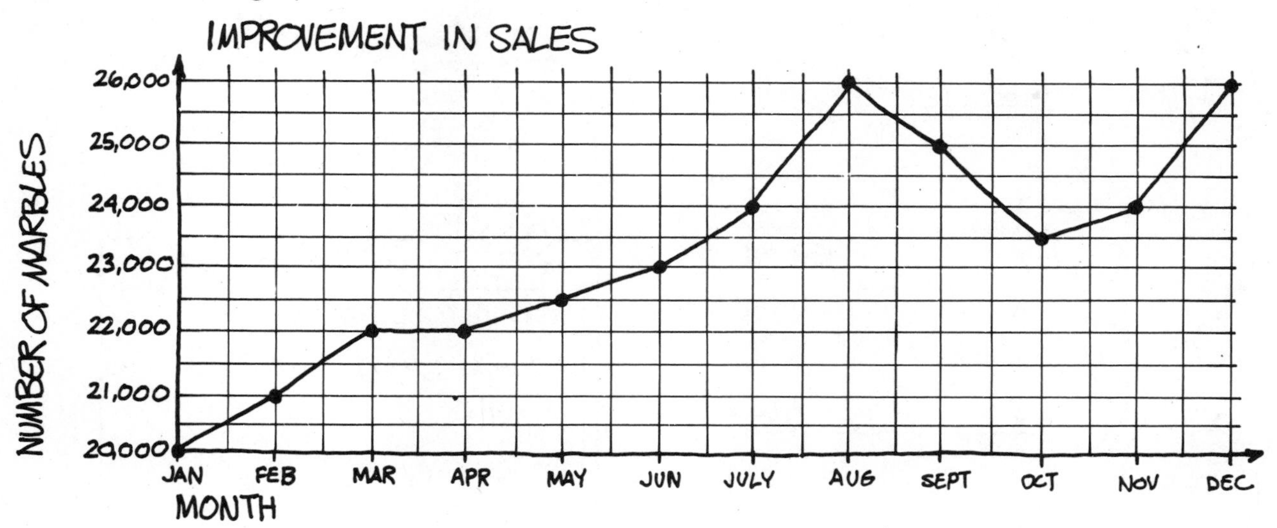

Redraw the graph on the grid below.

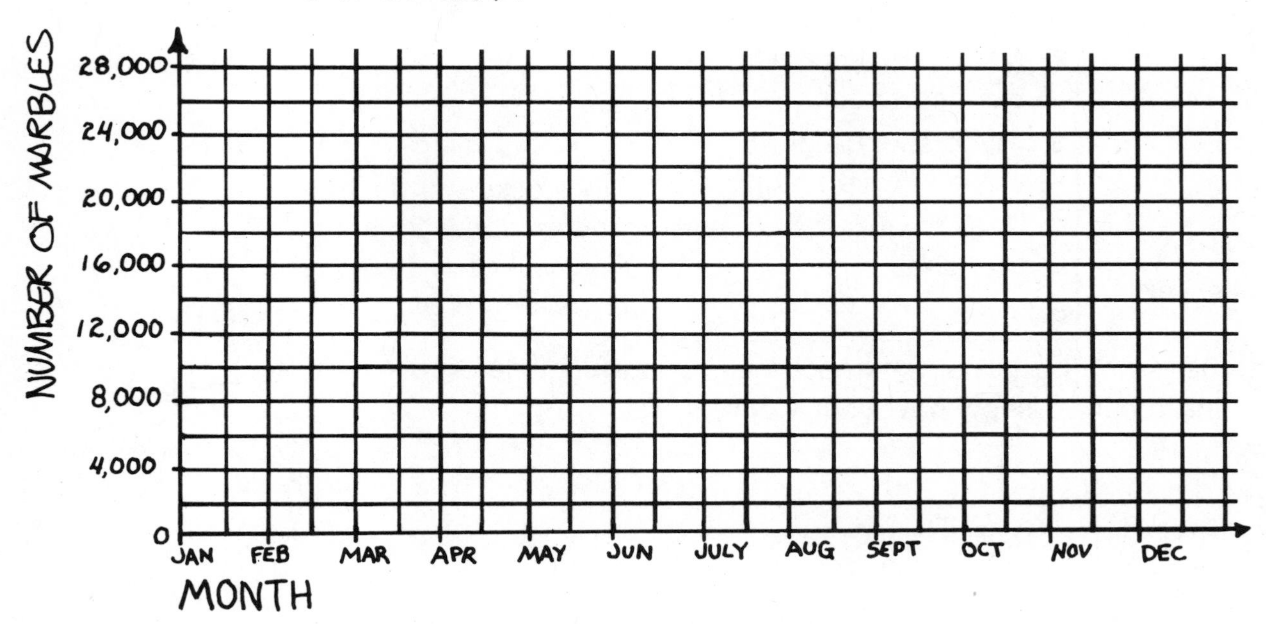

1. Which graph would you use to show big changes in sales? ____________
2. Which graph would you use to show steady sales? ____________
3. Describe three ways the graphs differ.

____

____

____

Name ______________________

ALIEN WORLD

You can create your own alien world. Here's how.

ingredient	amount
water	30 mL
salt	30 mL
liquid bluing	30 mL
ammonia	7 mL

USE THIS EQUIPMENT

MIAMI BEACH

equipment
measuring beaker
pieces of sponge or small stones
pie plate
mixing bowl
wooden spoon

directions
1. Measure ingredients into mixing bowl.
2. Mix ingredients with wooden spoon.
3. Place sponge or small stones in pie plate.
4. Pour mixture on sponge or stones.
5. Wait.

Alien rocks will begin to grow in about 24 hours. They will continue to grow for several days.

Find the height of the tallest rock each day. Use a centimeter ruler. Make a line graph to show how your world grows.

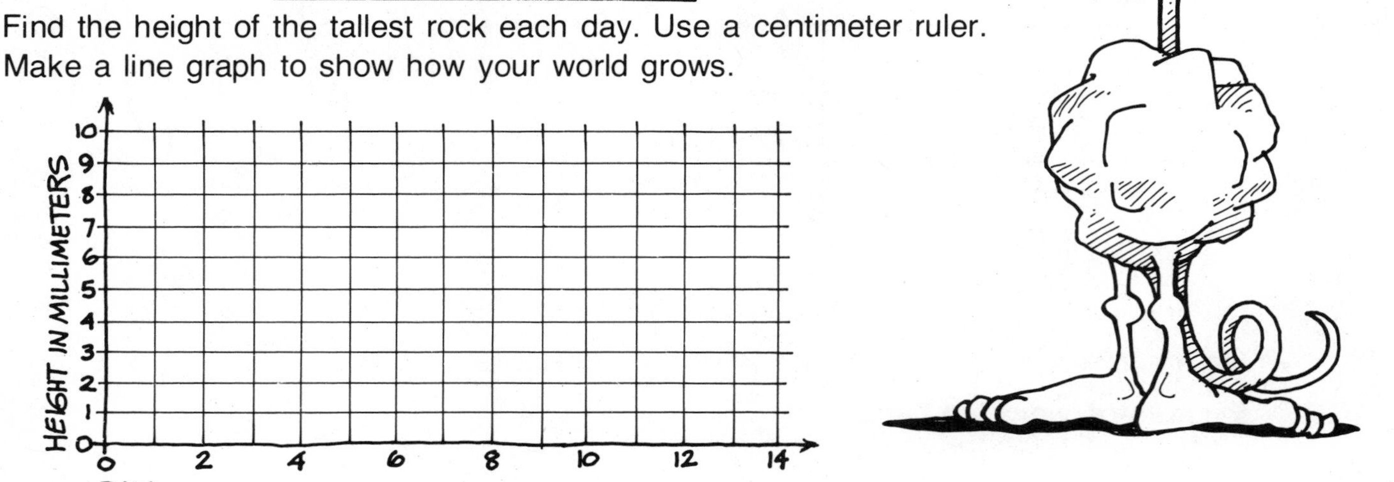

Each day find the height of the tallest rock. Use a centimeter ruler. Make a line graph to show how your world grows.

Name ______________________

CRICKET THERMOMETERS

Crickets notice temperature — at least their chirps change with the temperature. Some people say that the number of times a cricket chirps in one minute can actually be used to find the temperature. See what you think.

The following line graph shows some data about cricket chirping.

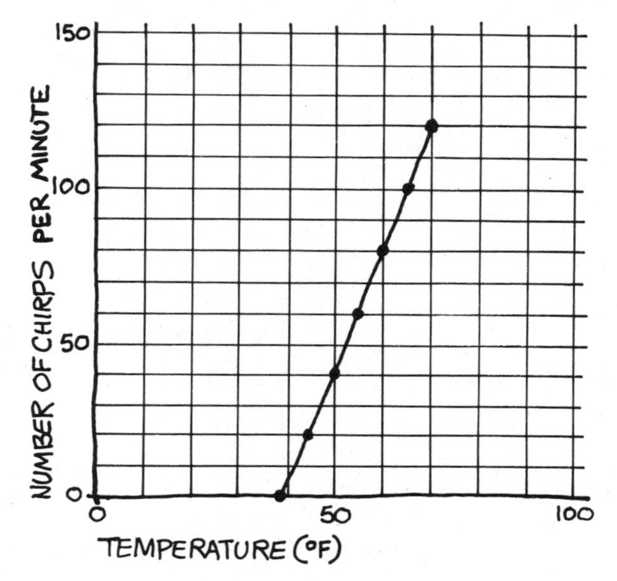

Use the line graph to help answer these questions about cricket chirping.

1. Suppose a cricket chirps 80 times in one minute. What is the temperature?

   ____

2. Suppose the temperature is 70°F. How many times a minute would you expect a cricket to chirp? ____________
3. At what temperature do you think crickets stop chirping? ____________
4. Suppose a cricket chirps 70 times in one minute. What is the temperature?

   ____

5. Suppose a cricket chirps 125 times in one minute. Would you expect the temperature to be warmer or colder than 70°F? ____________
6. Suppose the temperature is 87°F. About how many times a minute would you expect a cricket to chirp? ____________
7. As the temperature gets warmer, what happens to cricket chirps?

   ____

8. In the following equation, T stands for temperature and c stands for number of chirps. Some people say that the equation shows how cricket chirping and temperature are related. Do you agree? ____________

$$T = \frac{1}{4}c + 40$$

Name ____________________

HOW DO YOU FEEL TODAY?

Do you have the blues?
Are you tripping over your own two feet?
Have you forgotten how to add two plus two?
It could be that your biorhythms are at low points.

MONDAY 13

Complete the following to find your biorhythms.

1. 365

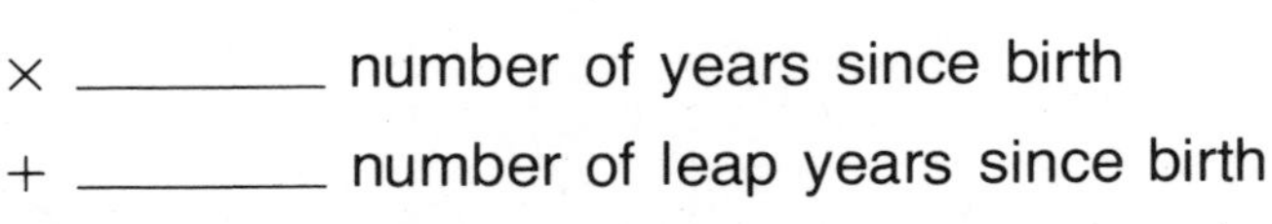

× ________ number of years since birth

\+ ________ number of leap years since birth

\+ or − ________ number of days between birthday and today (add if your birthday was earlier this year, subtract if not)

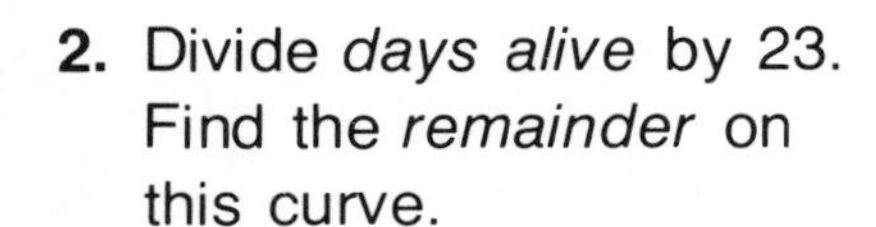

2. Divide *days alive* by 23. Find the *remainder* on this curve.

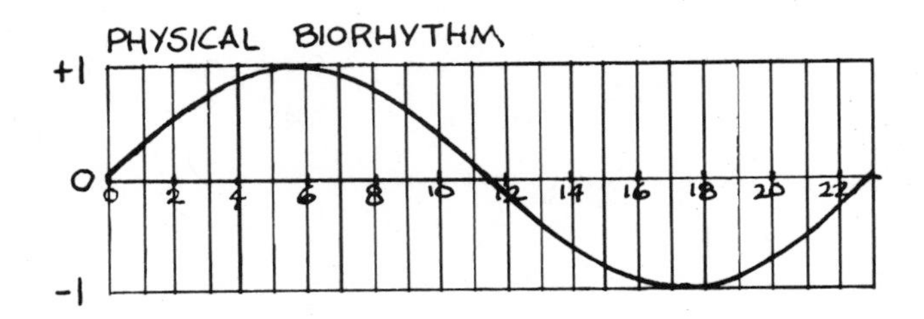

3. Divide *days alive* by 28. Find the *remainder* on this curve.

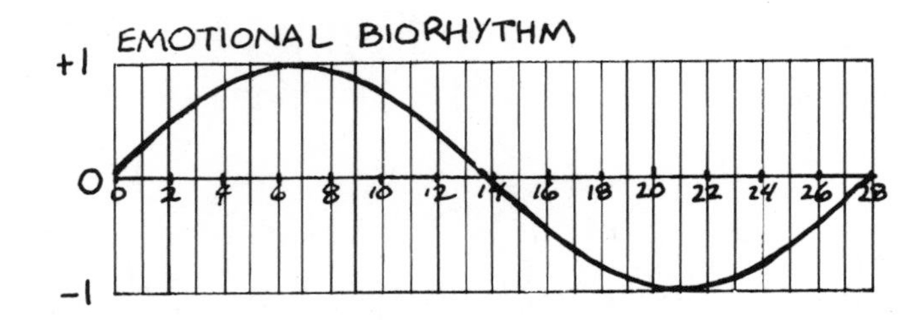

4. Divide *days alive* by 33. Find the *remainder* on this curve.

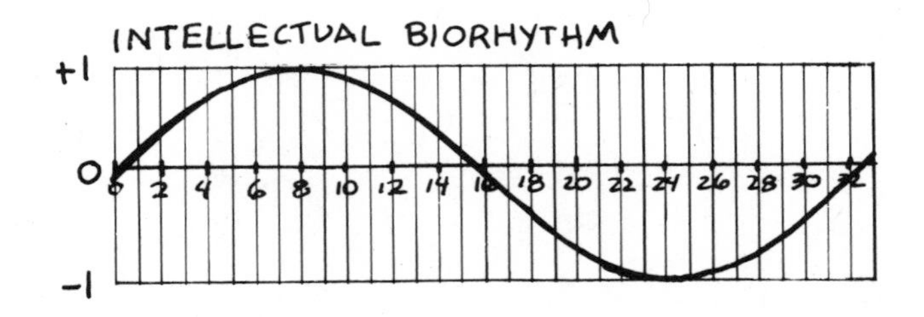

The closer your points are to +1, the better you may feel.
The closer your points are to −1, the worse you may feel.

Name ____________________

PHONY GRAPHS

Line graphs help show changes or trends. On a line graph, the points show the data that is given. The lines connecting the points help you estimate where other points might be.

A line graph is *not* always the right kind of graph to use. For example, the facts from *only one* of the following tables are suitable for a line graph.

TELEPHONES IN THE UNITED STATES BY YEAR									
Year	1900	1910	1920	1930	1940	1950	1960	1970	1980
Number (millions)	3	$8\frac{1}{2}$	13	21	23	43	72	120	$155\frac{1}{2}$

TELEPHONES IN THE UNITED STATES BY REGION						
Region	East	South-east	Mid-west	South-west	West	Alaska & Hawaii
Number (millions)	40	27	$45\frac{1}{2}$	16	26	1

1. Which table do you think gives facts that are suitable for a line graph?

____

2. Give a reason for your answer to **1.**

____

____

3. On a separate piece of graph paper, draw a line graph showing the facts given in the table you named for **1.** Attach your graph to this worksheet.

4. Describe a change or trend that your graph shows.

____

Name ______________________

SCATTER DIAGRAMS

Scatter diagrams use points to show pairs of values. They may help show whether or not there is a trend or relationship between the values. Scatter diagrams are different from line graphs because the points are *not* connected and because there may be several points for given values. The following scatter diagram shows the results of a study about Bummer Burgers.

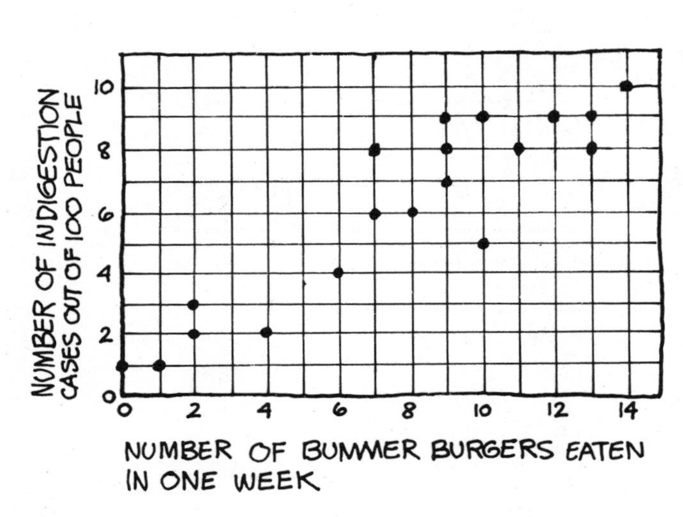

Answer each question about the scatter diagram.

1. How many points are on the scatter diagram? ______________
2. How many points show the results of people eating two Bummer Burgers in one week? ______________
3. How many points show the results of people eating nine Bummer Burgers in one week? ______________
4. Suppose 100 people eat 13 Bummer Burgers each in one week. About how many cases of indigestion would you expect, one, five, or ten? ______________
5. Do you think there is a relationship between the number of Bummer Burgers a person eats and the chances that person will get indigestion? ______________ Explain your answer. ______________________________

____

Name ______________________

STEPS FOR MAKING A LINE GRAPH

STEP 1 FIND THE RANGE IN VALUES

There are two sets of values.

	first set	second set
What units are used?	________	________
What is the greatest value?	________	________
What is the least value?	________	________

STEP 2 DETERMINE A SCALE

Start with the horizontal (side-to-side) scale.

Let one unit on the graph paper = 1 unit of the values you are graphing.

Will the greatest value fit on the graph? ________

If not, change the scale and try again.

Repeat the process for the vertical (up-and-down) scale.

Try to use the same scale for both sets of values.

Next, try plotting several pairs of values. Check several points with close values. Will their positions be different? ________

STEP 3 LABEL THE GRAPH

Mark each unit across the bottom and along the side of the graph.

Label the marks by the units they represent.

STEP 4 PLOT THE POINTS AND CONNECT THEM

Plot a point for each pair of values.

Which item of a pair is indicated by the horizontal scale? ________

Which item of a pair is indicated by the vertical scale? ________

How many points will you plot?

Connect the points with straight lines from left to right.

STEP 5 GIVE THE GRAPH A TITLE

What is your graph about? ________________________

____

Name ______________________

TABLE IT

Complete tables showing the information from each graph.

1.

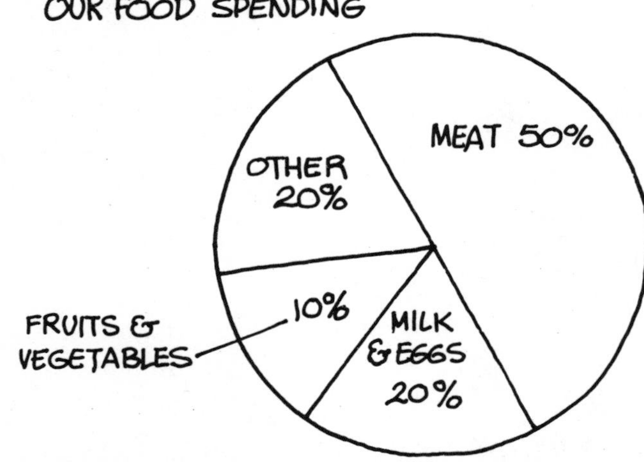

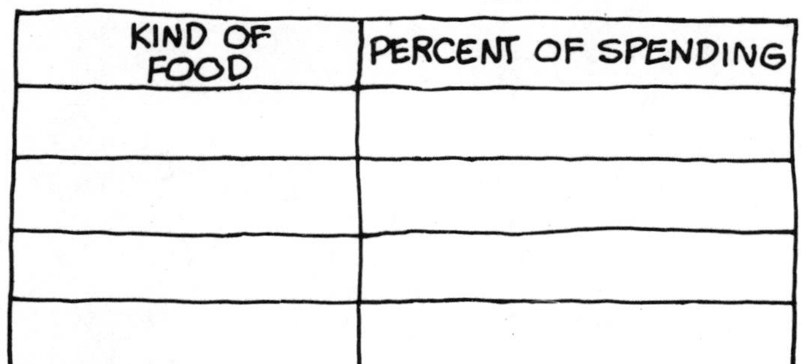

KIND OF FOOD	PERCENT OF SPENDING

2.

CALORIES IN FOOD (1 SERVING)

EACH SYMBOL = 50 CALORIES

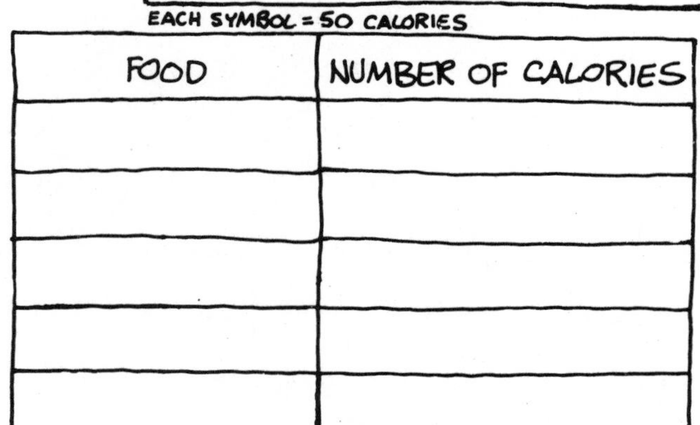

FOOD	NUMBER OF CALORIES

3.

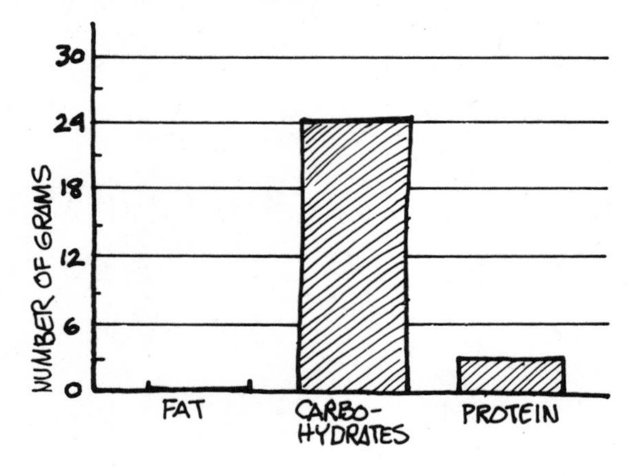

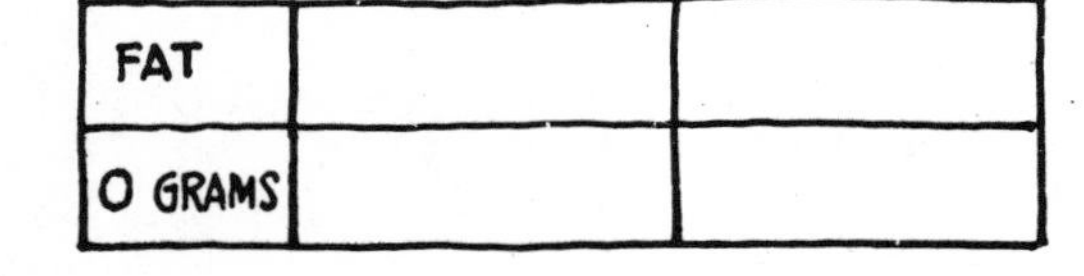

FAT		
0 GRAMS		

4.

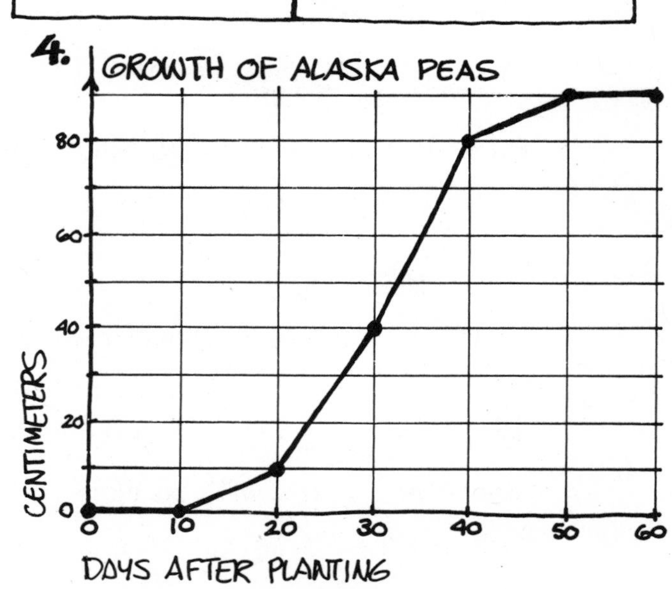

DAYS AFTER PLANTING							
HEIGHT IN CENTIMETERS							

Name ______________________

MIX AND MATCH

Each graph below answers one or more of these questions. Write the numbers of the questions next to the correct graph. Then answer the questions.

1. Which month had the coldest temperature? ______________
2. Which months had less than 6 cm of rain? ______________
3. How much money was spent for movies? ______________
4. How many movies did they see in all? ______________
5. Was there more rain in May or November? ______________
6. Was more money spent on movies or records? ______________
7. How did the temperature change from January to July? ______________
8. How much money was spent in all? ______________

______ MOVIES WE SAW LAST YEAR

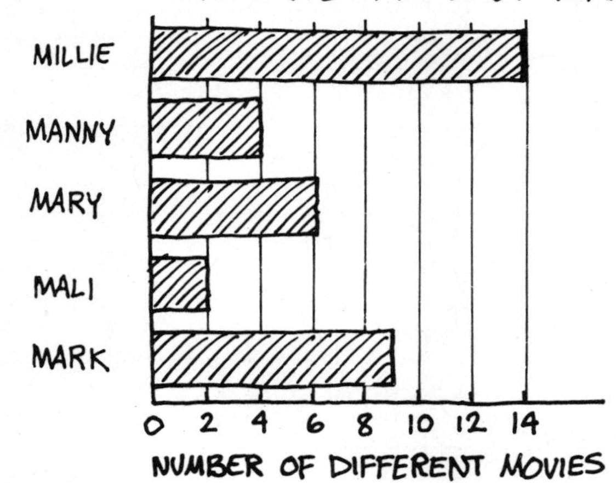

______ HOW I SPENT MY MONEY LAST YEAR

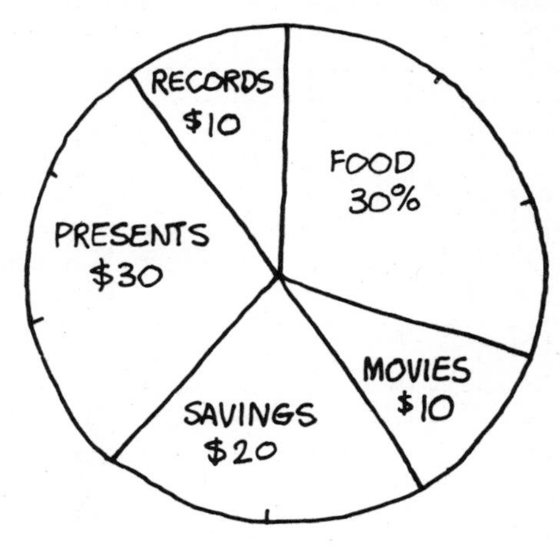

______ TEMPERATURE CHANGES LAST YEAR

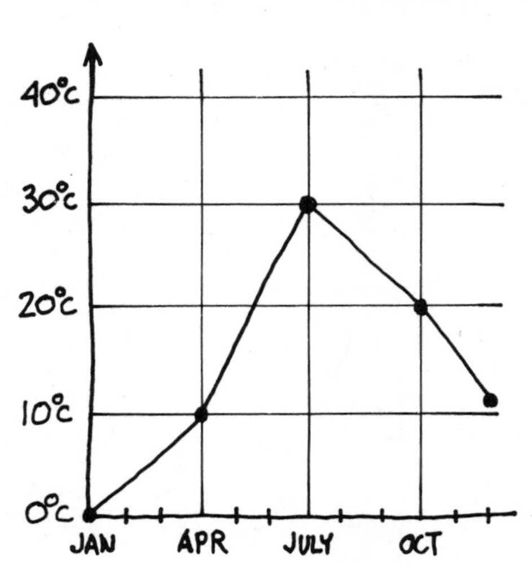

______ RAINFALL LAST YEAR

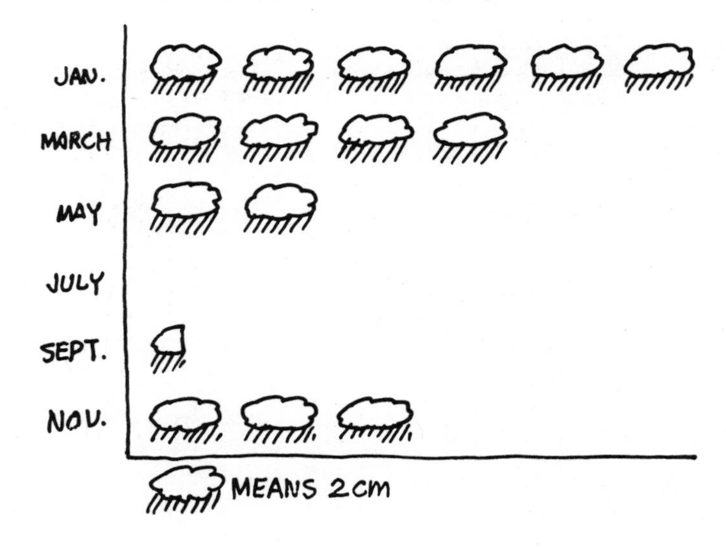

Name ______________________________

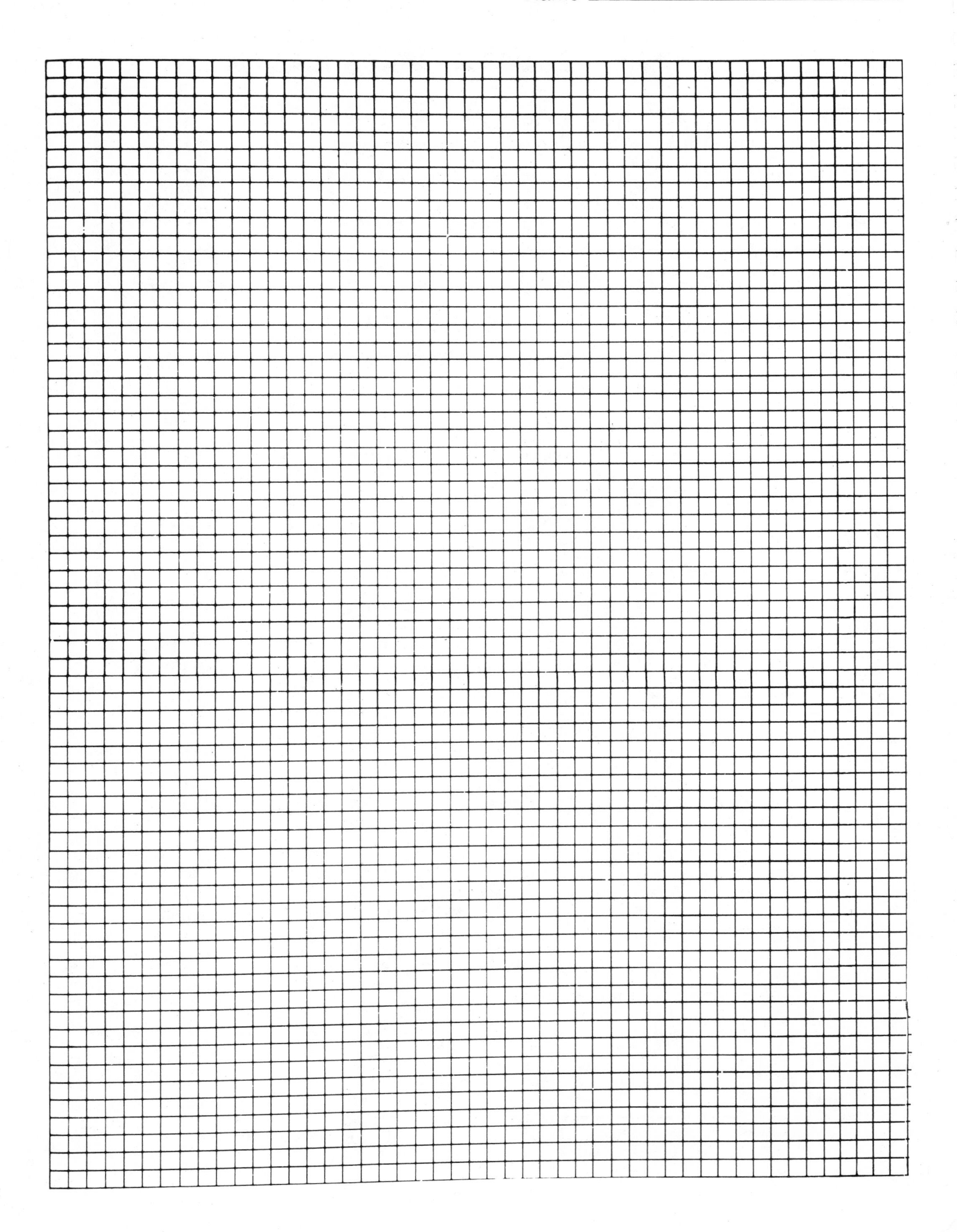

Name ______________________________

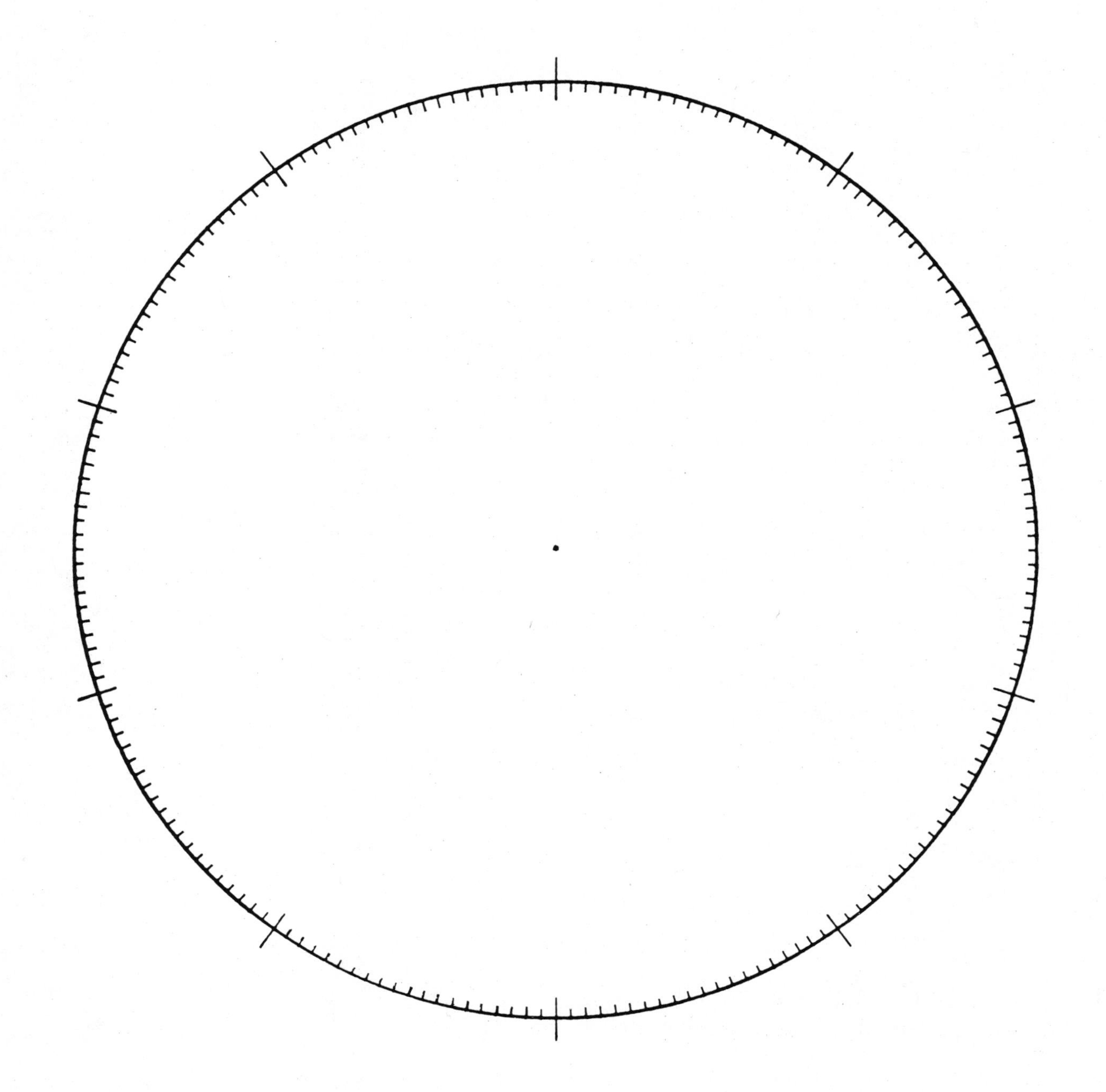

Name ______________________

HOW TO USE A PROTRACTOR

To draw an angle on a circle you use a tool called a protractor. Most protractors have two scales like the one shown here. To draw an angle using a protractor follow these steps.

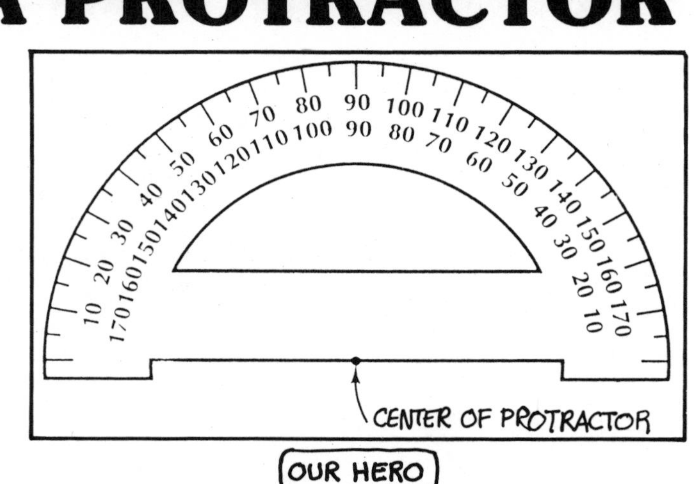

- Place the center of the protractor on the center of the circle. Line up the edge of the protractor with a radius of the circle.

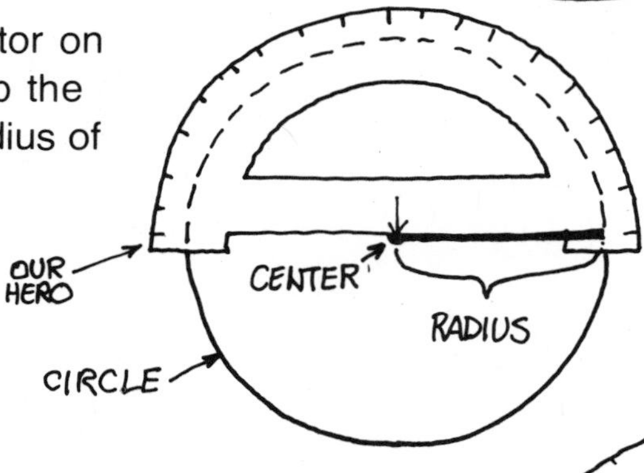

- Read the measure of the angle you want to draw by looking on the scale.

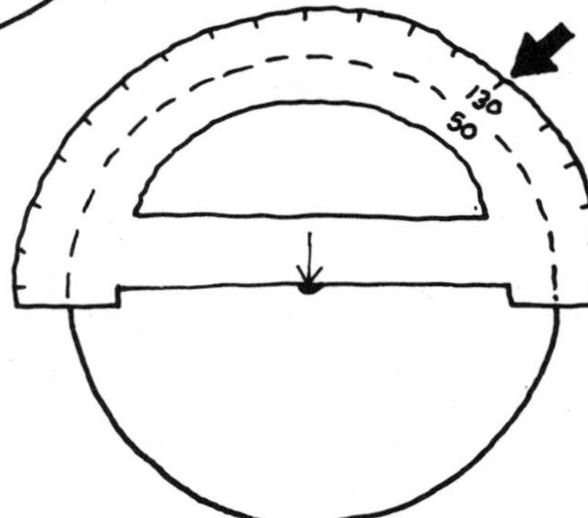

- Draw a point by the mark on the scale. Connect the point to the center of the circle.

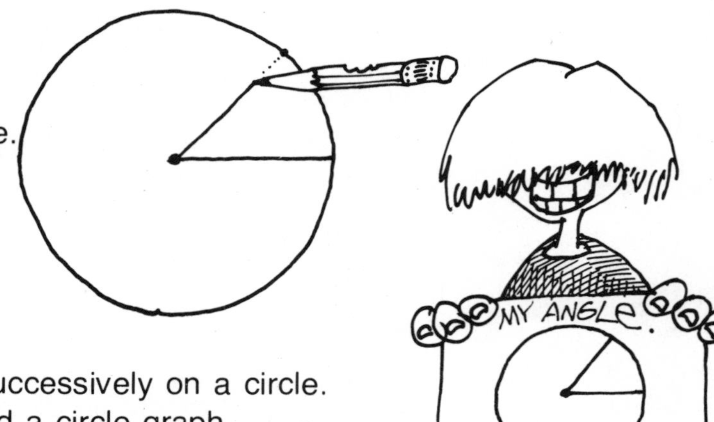

Use a protractor to draw these angles successively on a circle. When you finish, you will have completed a circle graph.

1. 90° **2.** 42° **3.** 30° **4.** 75° **5.** 105°

ANSWERS

page 1

1. 57,200,000
2. 8
3. 13,200,000
4. less than 5 or more than 10.
5. 6

page 2

1. $1.50, $0.40, $1.90, $0.12, $2.02
2. $1.80, $0.35, $2.15, $0.13, $2.28
3. $0.35, $0.35, $0.70, $0.05, $0.75
4. $1.05, $0.40, $0.60, $2.05, $0.12, $2.17

page 3

1. 11 km
2. $0.10 for each km
3. $1.10
4. $2.20
5. $3.30
6. $6.60
7. $13.20
8. 111 km
9. $1.20
10. $133.20

page 4

1. $2\frac{1}{2}$ hours
2. 40 minutes
3. 15 minutes
4. peas
5. rice, pudding, and chicken
6. roast chicken, 5:00 P.M.; rice, 6:50 P.M.; sweet potatoes, 7:00 P.M.; peas, 7:20 P.M.; chicken gravy, 7:15 P.M.; chocolate pudding, 6:40 P.M.
7. 4:40 P.M.

page 5

1. 9, $1.55
2. 8, $0.80
3. 4, $3.40
4. 3, $3.05
5. 4, $2.20
6. 1, $7.45
7. 5, $1.50
8. 1, $9.05
9. 18
10. $70.20
11. 5
12. $62.75

page 6

1. pencils, 68; pens, 80; 3-ring notebooks, 8; folders, 24; spiral notebooks, 20; assignment books, 2
2. 202
3. about 5
4. Any answer within the given range of values is acceptable. pencils, 330-350; pens, 390-410; 3-ring notebooks, 37-43; folders, 115-125; spiral notebooks, 90-110; assignment books, 8-12

page 7

Times in each column increase by 1 hr successively.

1. 2:00
2. Pacific
3. 4:00
4. 10:32

page 8

1. Franklin Mint's Monthly Expenses
2. rent
3. other
4. clothing
5. $50
6. $50
7. $75
8. $625
9. Answers may vary.

page 9

1. hair dryer, 1000 watts; refrigerator, 650 watts; color TV, 300 watts; record player, 100 watts
2. hair dryer, 2000 watts; refrigerator, 1300 watts; color TV, 600 watts; electric stove, 3000 watts
3. 100
4. hair dryer: 2000 4000 6000 12,000 24,000
   100 watt light bulb: 200 400 600 1200 2400
   electric stove: 3000 6000 9000 18,000 36,000
   color TV: 600 1200 1800 3600 7200
   radio: 100 200 300 600 1200
   record player: 200 400 600 1200 2400
   refrigerator: 1300 2600 3900 7800 15,600
5. 43 cents
6. less than 1 cent

page 10

1. 18
2. 12
3. 22
4. 10
5. 30
6. See the graph.

WEATHER IN APRIL

7. $\frac{8}{30}$ or $\frac{4}{15}$
8. $\frac{4}{30}$ or $\frac{2}{15}$
9. $\frac{12}{30}$ or $\frac{6}{15}$ or $\frac{2}{5}$
10. $\frac{4}{30}$ or $\frac{2}{15}$

page 11

1. 2 stamps
2. 12
3. 10
4. 3
5. 52
6. 6
7. 31
8. 2
9. not enough information
10. 8

page 12

1. 10 records
2. 5 records
3. 50
4. 20
5. 35
6. not enough information
7. 180
8. 160
9. 45
10. jazz
11. not enough information
12. yes (10 of them)

page 13

order may vary.

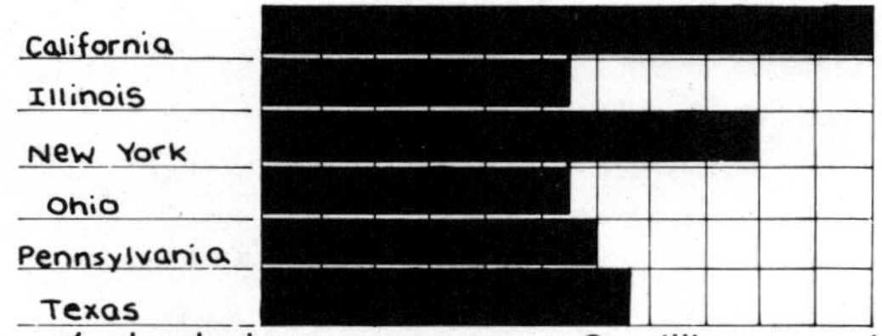

1 shaded square means 2 million people

1. 1 million people
2. Illinois
3. 11 million
4. 11 million
5. They appear to be the same.

page 14

Answers will vary depending on the data used.

page 15

1. percentage of sugar in certain foods
2. chocolate bar
3. cola
4. ketchup
5. 15.5%
6. 42.6
7. cola, peanut butter, crackers, ice cream, ketchup, chocolate cake, chocolate bar
8. start at 0 and end at 60

page 16

1. The third graph shows the facts correctly.
2. The first graph shows the facts correctly.
3. The second graph shows the facts correctly.

page 17

1. scale is too spread out, exaggerating differences
2. measurements in scale not consistent
3. scale does not start at 0
4. scale is too spread out, exaggerating differences

page 18

Answers may vary.

1. table
2. graph
3. table
4. graph
5. either
6. table
7. graph
8. table

page 19

1. no
2. yes
3. yes
4. yes
5. no
6. yes
7. yes
8. no

page 20

1. between Larch and Maple
2. before Alder
3. between Larch and Maple
4. 21
5. Answers may vary. Be sure that the reasons given accurately represent the information from the table.

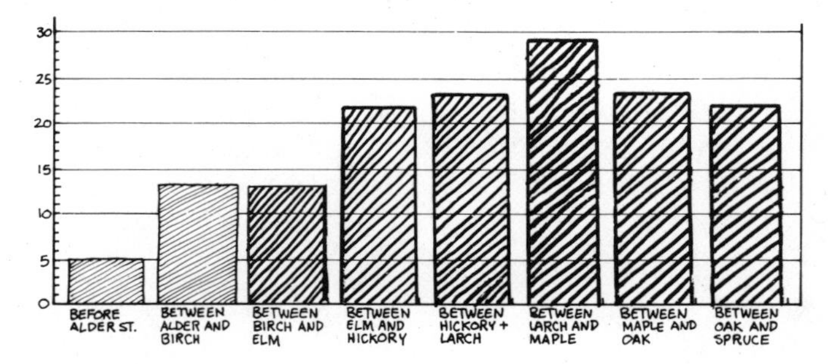

page 21

1.

97	I
98	
99	I
100	I
101	IIII
102	III
103	HHT III
104	III
105	HHT I
106	II
107	
108	I

2. 8

3. 97 cm, 108 cm
4. and 5. See the graph.

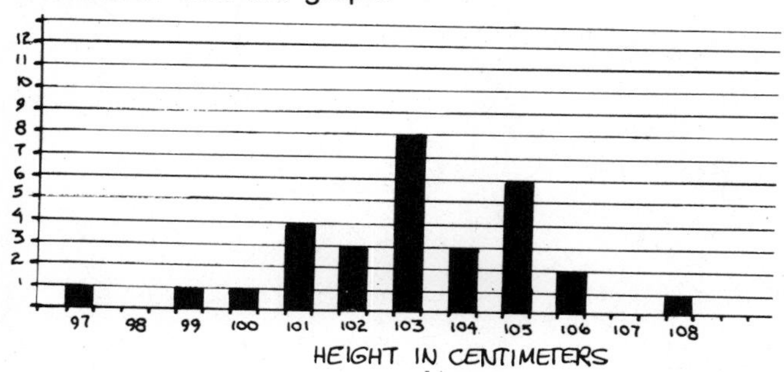

page 22

1. Bette A. Million
2. Willy Maykett
3. Ben D. Rules
4. Anita Win

page 23

1.

Score	Number
20-25	1
26-30	3
31-35	6
36-40	5
41-45	2
46-50	1

2.

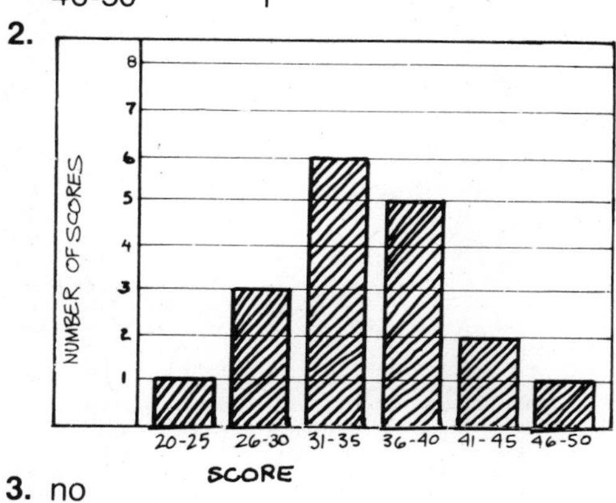

3. no

page 24

Answers will vary depending on the data used.

page 25

1. Jolene Simpson's budget
2. 5
3. on rent
4. food, 20¢; rent, 50¢; clothing, 15¢; entertainment, 5¢; other, 10¢
5. $200
6. $500

page 26

1. The first graph shows the facts correctly.
2. The third graph shows the facts correctly.
3. The second graph shows the facts correctly.

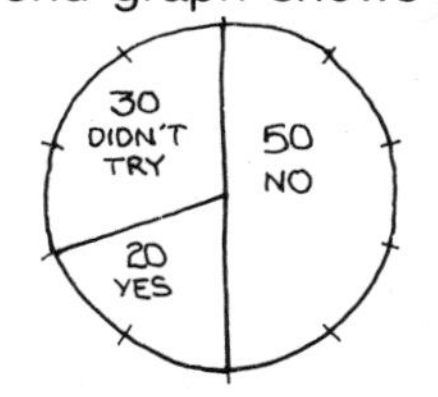

page 27

1.

rubber plants	$\frac{11}{85}$,	0.13, 13%
snake plants	$\frac{8}{85}$,	0.09, 9%
geraniums	$\frac{13}{85}$,	0.15, 15%
begonias	$\frac{10}{85}$,	0.12, 12%
cacti	$\frac{16}{85}$,	0.19, 19%
jade plants	$\frac{6}{85}$,	0.07, 7%
palm trees	$\frac{11}{85}$,	0.13, 13%
other	$\frac{10}{85}$,	0.12, 12%

2.

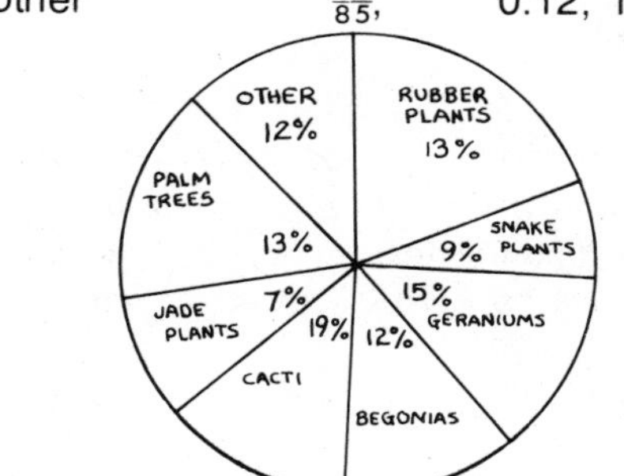

3. cacti, geraniums, rubber plants and palm trees, begonias and other, snake plants, jade plants

page 28

1.

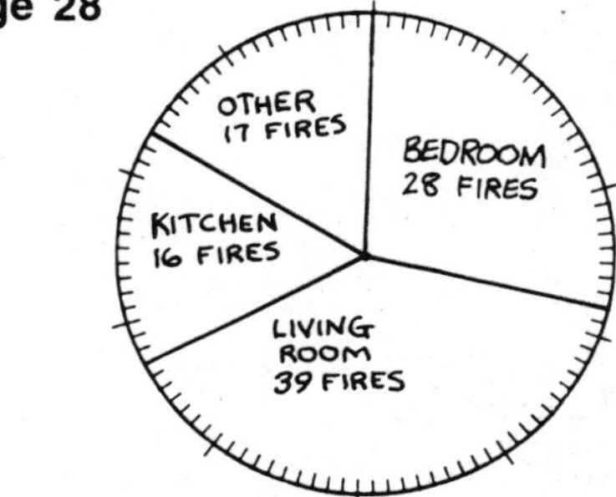

2. the living room
3. 2
4. living room and bedrooms

page 29

1. 34, 100
2. 14, 41
3. 20, 59
4. 18, 53
5. 16, 47
6. 5, 15
7. 9, 26
8. 13, 38
9. 7, 21
10.

11. 300
12. 520
13. 760
14. 420

page 30

1. Number of half-hours for TV shows
2. news, movies, reruns, talk shows, sit-coms, drama, game shows, other
3. 7 (8 if other is included)
4. 14
5. 2
6. movies
7. game shows
8. sit-coms
9. Probably the record is about a weekday because there is no category for sports. Make a breakdown of the TV shows for weekdays, Saturdays, and Sundays from information given in a newspaper TV section or in a TV guide. Then, compare the results to the facts in Terry's graph.

page 31

1. rent, \$125; gas, \$25; electric, \$6.50; water, \$6; rent and utilities, \$162.50
2. rent and utilities, \$162.50; food, \$100; car, \$225; clothing, \$75; entertainment, \$25; other, \$50; total, \$637.50
3.

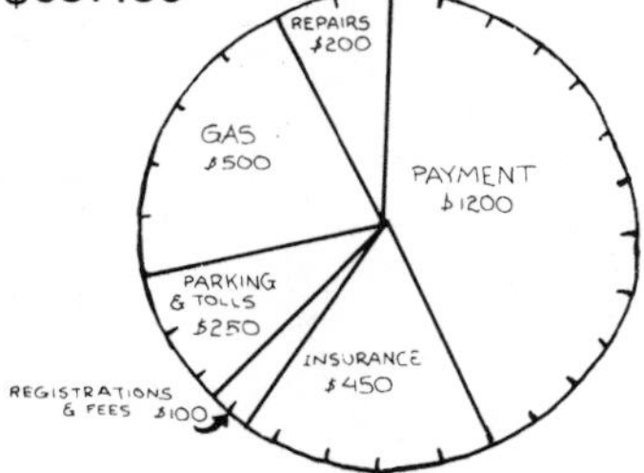

page 32

Answers will vary depending on the data used.

page 33

1. People going in and out of a store.
2. 1:00 PM
3. about 3:00 PM
4. 2
5. about 10
6. 22
7. 2

page 34

1. Kansas City
2. Miami
3. Eureka
4. Honolulu
5. Fairbanks
6. New York

page 35

1. 11.00 sec
2. 11.40 sec
3. from 1948-1952, decreased;
   from 1952-1956, stayed the same;
   from 1956-1960, decreased;
   from 1960-1964, increased;
   from 1964-1968, decreased;
   from 1968-1972, increased;
   from 1972-1976, decreased
4. 10.70 sec
5. Answers may vary.

page 36

1. from left to right the correct numbers are: 0.60, 0.64, 0.60, 0.60, 0.55, 0.50, 0.50
2.

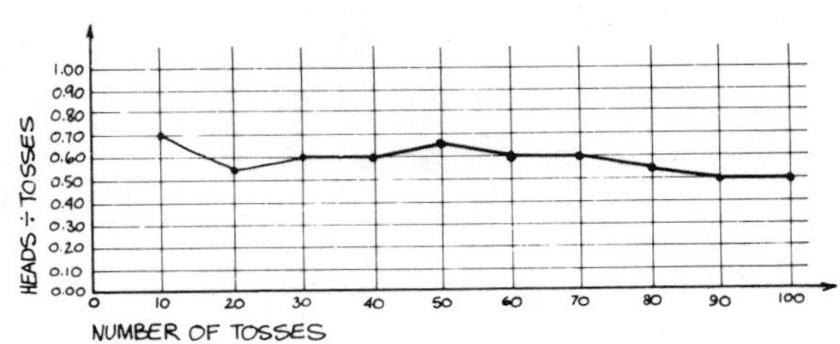

3. from 10-20, decreases
   from 20-30 increases
   from 30-40, stays the same
   from 40-50, increases
   from 50-60, decreases
   from 60-70, stays the same
   from 70-80, decreases
   from 80-90, decreases
   from 90-100, stays the same
4. about 0.50
5. about 55 altogether

page 37

1. the first graph
2. the second graph
3. Answers may vary. The titles differ, the vertical scales differ, and the shape of the lines differ.

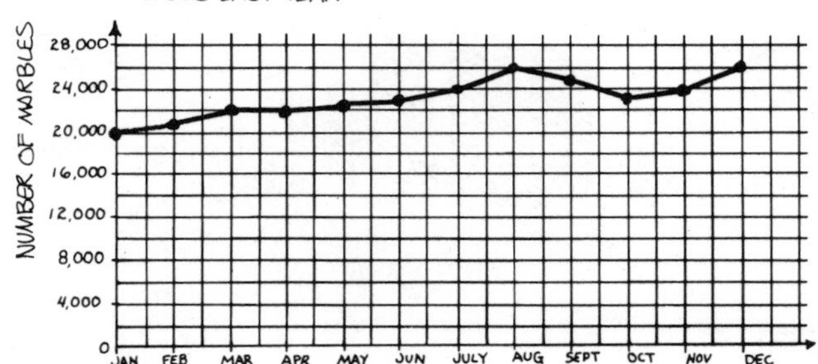

page 38

Answers will vary.

page 39

1. about 60°F
2. about 120
3. 40°F
4. between 55°F and 60°F (57.5°F)
5. warmer
6. more than 150 times (188 times)
7. They get faster.
8. Yes. The equation is correct.

page 40

Answers will vary.

page 41

1. the first table
2. Answers will vary. The first table gives information that shows a trend over time and permits interpolation.
3.

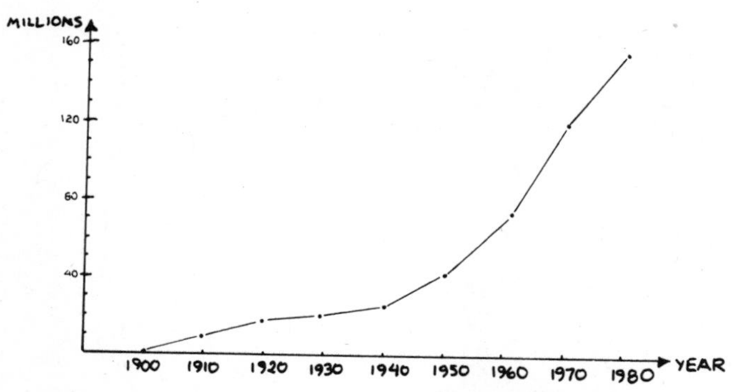

4. Answers will vary. The number of telephones increases as the years approach the present.

page 42

1. 19
2. 2
3. 3
4. ten
5. Answers will vary. The graph does suggest that a relationship exists.

page 43

Answers will vary depending on the data used.

page 44

1. Our Food Spending

KIND OF FOOD	PERCENT OF SPENDING
meat	50%
milk & eggs	20%
fruits & vegetables	10%
other	20%

2. Calories in Food (1 serving)

FOOD	NUMBER OF CALORIES
Yogurt	125
Pork Chop	300
Egg	75
Burger	300
Chicken	200

3. Nutritional Content of Krummy Korn

FAT	CARBOHYDRATES	PROTEIN
0 grams	24 grams	3 grams

4. Growth of Alaska Peas

DAYS AFTER PLANTING	HEIGHT IN CENTIMETERS
0	0
10	0
20	10
30	40
40	80
50	90
60	90

page 45

1. January
2. May, July, September
3. $10
4. 35
5. November
6. Same
7. It increased.
8. $100

Movies We Law Last Year: 4
How I Spent My Money Last Year: 3, 6, 8
Temperature Changes Last Year: 1, 7
Rainfall Last Year: 2, 5